Populations and Ecosystems

Developed at
The Lawrence Hall of Science,
University of California, Berkeley
Published and distributed by
Delta Education,
a member of the School Specialty Family

1558518
978-1-62571-788-7
Printing 1 — 11/2017
Webcrafters, Madision, WI

Table of Contents

Observations and Inferences

The world around us is filled with millions of sights, sounds, smells, textures, and tastes.

We take them in and process the information with our brains. "This rose smells sweet." "Ouch, the thorn is sharp!" "That barking dog must be scared." You are always making **observations** and **inferences**.

Look at this picture. What do you observe? Make a list of your observations in your science notebook. Observations provide factual information through the five senses: sight, touch, smell, hearing, and taste. Observations can be qualitative (describing a quality), using adjectives such as *small* or *shiny*, or quantitative (describing a quantity), using numbers such as *3 centimeters (cm)* or *20 seeds*.

Observations of these insects could include their color, body shape, and how many are clustered on the milkweed pod. Can you make any inferences about their diet?

Inferences are explanations or assumptions that people make based on their knowledge, experiences, or opinions. Inferences are different from observations. Two people could observe the same thing but make different inferences. "The grass is wet this morning" is an observation. If you did not observe the grass getting wet, you might infer "It rained last night" or "The sprinklers came on this morning."

Sometimes our inferences are limited by our ability to get facts. For example, when we observe a milkweed bug heading toward the **food** source in its **habitat,** we might infer that the bug feels hungry. But we cannot really know how it feels. We can know only what we observe, which is the direction in which it is moving.

Let's make some observations again, this time focusing on the difference between observation and inference. Analyze the picture of the bonobos below.

In your notebook, make a T-chart with one column labeled "Observations" and the other labeled "Inferences."

What are your observations of the bonobos?

Put each statement below in one of these columns.

1. There are two **individuals** of the same kind of animal.
2. These animals are in the wild.
3. One animal is mad at the other.
4. Both animals have open mouths and are looking toward each other.
5. Both animals are in the same tree.
6. The smaller animal is leaning backward.
7. The smaller animal is the child of the larger animal.
8. The smaller animal is scared.

Humans are very social, emotional animals. We have a great capacity for empathy—the ability to understand and share the feelings of another. Sometimes we make inferences that we think explain another animal's actions, but we have no way to check these inferences. For example, we might assume that the larger bonobo is mad at the smaller one. However, there could be another explanation for their actions that we don't know about. The assumption that other animals think like humans is called anthropomorphism.

Scientific thinking starts with detailed, accurate observations. Then, scientists evaluate their data to make inferences that can help them understand the phenomena they have observed. Just like scientists, when we make inferences, we need to be clear that they are not observations. Each time you make an observation throughout this course, think about whether you are actually making an inference. Challenge yourself to stick to the facts!

Think Questions

1. Look in your notebook at your first observations of the milkweed bugs. Were any of your observations actually inferences? If so, mark them with an *I*.
2. Can you explain your inferences with observations? Replace your inferences with observational statements.

Scientists, including science students, rely on tools like microscopes to extend their senses and take measurements when making observations.

Milkweed plants provide both a habitat and a food source for milkweed bugs. The insects lay their eggs on the plant and consume the nectar inside the ripening seeds.

Milkweed Bugs

In class, you started making observations of a pair of milkweed bugs. This small population will soon increase. What can you look forward to seeing?

Milkweed bugs are easily recognized as insects. They have the same structures as most other insects: six legs, three body parts (head, thorax, and abdomen), and two antennae.

Milkweed bugs are true bugs. They do not have mouths for biting and chewing food. Instead, they have a **proboscis**, which is a tube-like beak for sucking fluids.

In nature, the milkweed bug uses its proboscis to pierce and suck nutrients from the seeds of the milkweed plant. The bugs in your classroom, however, have been bred to feed exclusively on raw, shelled sunflower seeds. Your classroom milkweed bugs insert their long beaks into sunflower seeds to suck out the oils and other nutrients.

Milkweed-Bug Life Cycle

Eggs

1st instar

4-7 days

2nd instar

2-3 days

3rd instar

3-5 days

4th instar

5-10 days

10-15 days

5th instar

10-15 days

Adult

Tiny milkweed bugs hatch from bright orange eggs and gradually mature through several molts to adulthood.

How Do Milkweed Bugs Grow?

Milkweed bugs start life as tiny eggs. The eggs hatch about a week after being laid. The new bugs are not much bigger than the period at the end of this sentence. If you look at a newly hatched milkweed bug under a microscope, you will see a tiny insect, with six legs, three body parts, and two antennae. You will also see that it has a tough outer covering called an **exoskeleton** to protect it.

The exoskeleton is not flexible, so the tiny bug cannot grow while the exoskeleton is in place. After a few days, the immature milkweed bug, called a **nymph**, bursts its exoskeleton and sheds it. The nymph is then protected by a new exoskeleton. The new exoskeleton is moist and flexible. The bug pumps itself up, growing to twice its original size within minutes. Within a few hours, the exoskeleton hardens, and the larger nymph goes about its business of eating and growing.

About a week after the bugs hatch, brittle transparent exoskeletons with black squiggly legs start to accumulate in the bottom of the habitat. Shedding the exoskeleton in order to grow is called **molting**. Just after molting, the bug is creamy yellow with bright red legs and antennae. Within a few hours, the body turns orange, and the legs and antennae become black again.

Like other milkweed feeders, the milkweed bug gets some defensive poisons from nectar, making it unattractive to predators.

The milkweed bug molts five times before it becomes a fully mature adult. With each molt, the body shape changes, and the bug develops more dark body markings. Soon, wings start to form. Each nymph stage is called an **instar**. The first instar is the newly hatched bug. The fifth instar is the one just before adulthood.

This gradual maturing of an insect is called **incomplete metamorphosis**. The bug gets steadily bigger and more complete until the last molt reveals the adult. The process from egg to adult takes 5–8 weeks, depending, to a large extent, on the temperature. A week or more after reaching adulthood, the bugs will mate, and the female will lay eggs. In a room that is a comfortable temperature for humans, the eggs will hatch in about a week, changing from lemon yellow to tangerine orange as they mature. About half the bugs that hatch will be males and half females. The life cycle from egg to egg is about 2 months.

Take Note

Look at your milkweed-bug observation log. What evidence of bug growth have you observed? Which of the instars have you observed?

What Are the Mating Habits of Milkweed Bugs?

You can easily observe mating, as the two mating bugs remain attached end-to-end for several hours. You can tell the female adults from the male adults by the body markings on the ventral (belly) side of the bugs. The tip of the abdomen is black on both sexes. Next comes a solid orange segment (with tiny black dots at the edges). If the next two segments after the orange segment are solid black bands, it is male. If the next segment after the orange segment is orange with two large black spots, followed by a solid black band, it is female.

Males tend to be smaller than females. When you see a mating pair, observe closely to see if the female is noticeably larger than the male.

A few days after mating, the female starts laying clusters of 20 or more yellow eggs. The clusters are called **clutches**. A female might lay five or more clutches. In nature, the female lays her eggs in a ball of milkweed seed fluff or under a bit of bark for protection. In your classroom habitat, she will usually lay them in the polyester wool.

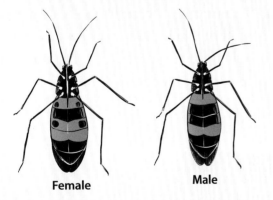

Milkweed Bugs

Female   Male

Size and segment coloration are slightly different in the female and male milkweed bug.

After mating and laying eggs, adult milkweed bugs might live another 2 months in a kind of retirement. The life span of milkweed bugs in a sheltered habitat bag in a classroom is about 4 months. In the wild, they probably do not live that long.

In captivity, a milkweed-bug **population** will reproduce one generation after another. If the 2-month life cycle continues, six generations can be produced in 1 year! That's potentially a lot of milkweed bugs.

In the wild, however, milkweed bugs stop reproducing in the fall when the weather gets cold and the milkweed plants die. The adult bugs that have not reproduced find a place to hibernate. Even though they have natural antifreeze in their bodies, winter takes its toll on the population. However, at least a few bugs always live until spring. And when a male and a female survivor meet on a fine spring day, they naturally continue the process of building up the population.

Take Note

Look at your milkweed-bug observation log. When did you first observe evidence of mating in your habitat? What evidence have you observed?

Milkweed bugs attached end-to-end are mating. The female will soon lay a clutch of eggs in a protected spot, like a cloud of seed fluff.

Milkweed bugs jostle for feeding sites on a seed pod.

Oncopeltus fasciatus

For purposes of classification, all life on Earth is divided into three domains— Bacteria, Archaea, and Eukaryota. Within Eukaryota, scientists have classified **organisms** into kingdoms. The kingdoms are divided into major groups called phyla. Each phylum is divided into classes. Each class is divided into orders. Each order is divided into families. Each family is divided into genera, and each genus into **species**. A species is a basic category or kind of organism. A species consists of individuals that are similar in structure and that can breed to produce offspring.

The milkweed bug studied in this course is a member of the phylum Arthropoda, the class Insecta, the order Hemiptera (true bugs), the family Lygaeidae (seed bugs), the genus *Oncopeltus*, and the species *fasciatus*. Its common name is large milkweed bug, and its scientific name is *Oncopeltus fasciatus*. A close relative of the large milkweed bug is *Oncopeltus sexmaculatus*, the six-spotted milkweed bug (same genus, different species). A more distant relative of these two bugs is *Lygaeus kalmii*, the small milkweed bug (same family, but different genus and species).

Think Questions

1. **Look at your milkweed bug habitat. On what plant structures of the milkweed plant do you think you would find wild milkweed bugs?**

2. **How can you tell if a milkweed bug has just molted?**

Life in a Community

You live in a community. But it might not be the *community* you think you live in. One meaning of *community* is people meeting, working, planning, and living together. But that's not what community means to an ecologist.

The Ecological Community

From a scientific point of view, your **community** holds populations of plants, animals, and other organisms that live and interact in an area. It includes only the populations living in a place, not the place itself. Your ecological community might include a population of rodents, several populations of trees, many populations of grasses, and hundreds of populations of insects. Most communities also include countless populations of microscopic organisms, such as algae, fungi, and bacteria.

Coral reefs are home to an amazing diversity of life. All the populations of aquatic plants, fish, and other marine organisms living together in an area make up a unique ecological community.

A community is described by the organisms living in it. Therefore, no two are exactly the same. The community of organisms on an island in Lake Superior is like, but not the same as, the community on the shore of Lake Superior near Marquette, Michigan. Both of those communities, however, are very different from the community of organisms on a coral reef off Key West, Florida.

Island	Lakeshore	Coral Reef
Sparrow	Sparrow	Algae
Crow	Crow	Sea grass
Osprey	Osprey	Sponge
Mosquito	Mosquito	Coral
Black fly	Black fly	Lobster
Badger	Deer	Moray eel
Skunk	Badger	Angelfish
Squirrel	Fox	Gobi
Mouse	Skunk	Nudibranch
Pine	Mouse	Snail
Spruce	Moose	Sea star
Oak	Pine	Octopus
Beech	Spruce	Sea urchin
Blueberry	Oak	Shark
Moss	Beech	Porpoise
Grass	Blueberry	Shrimp
	Grass	Barracuda

Interactions

The interactions among organisms also define a community. Every organism's life is connected to every other organism's life in some way.

Some interactions are obvious. A robin eats a worm, and the population of worms decreases by one. Other interactions are not as easy to see. How is an ash tree important for a robin? It is a safe place to build a nest. Without the protection provided by the tree, the robin population might decrease.

How is a robin important for an ash tree? When the robin dies, its body falls to the forest floor. Bacteria and fungi consume the remains. They recycle minerals from the robin's body into the **environment**. The ash tree benefits from the minerals returned to the soil. A healthier ash tree is more likely to grow larger and produce seeds to make new ash trees. The robin's minerals remain, recycled into the soil, and nourish the tree. The tree in turn provides more nesting sites for the next generation of robins. The death of one robin affects not just that robin, but the entire robin population.

This is just one peek into the complexity of a community. Interactions among organisms, together with their **abiotic** (nonliving) surroundings, are called an **ecosystem**.

Take Note

What organisms do you interact with in your community? Which organisms do you eat, and which ones eat you? Which organisms compete with you for food, and which ones provide shelter or comfort?

Think Questions

1. **List the organisms in your community.**
2. **What are some of the organisms in your community that you cannot see?**

Feeding relationships are one way that populations in a community interact. Robins and earthworms are predator and prey.

Ecoscenario Introductions

Can you imagine yourself swimming over a coral reef, or hiking among geysers and boiling pools? You can do both of these activities, and many more, in the United States.

Every environment on Earth has certain conditions that define it. The conditions of water, air, soil, and local geology are especially important. Because the United States extends from the Arctic to the tropics, the country has an amazing range of environmental conditions.

This environmental diversity supports an incredible diversity of organisms. Something is living wherever you look—on land, in water, on mountain peaks, on desert sands. An ecosystem is defined by the physical environment and the organisms that live there. Ecosystems that have similar environments and organisms are called **biomes**.

Ecoscenarios

In this course, you are going to learn about 11 ecosystems. They are located in national parks and preserves. The ecoscenarios in this article present 10 ecosystems and ecosystem issues. The next investigation introduces the 11th ecosystem for closer study.

Ecoscenario Locations

Arctic National
Wildlife Refuge

Yellowstone
National Park

Mono Lake

Monterey Bay National
Marine Sanctuary

Tallgrass Prairie
National Preserve

Sonoran Desert
National Monument

Delaware Water Gap
National Recreation Area

Monongahela
National Forest

Everglades
National Park

Florida Keys National
Marine Sanctuary

El Yunque
National Forest

The 11 US ecoscenarios we will explore span a stunning variety of environments and communities. Each location has a distinctive ecosystem.

An ecosystem service is a benefit that people gain from an ecosystem. Recreation such as snorkeling is an ecosystem service.

Each ecoscenario introduces the key organisms that populate it. Some organisms have a flexible lifestyle. They can survive in several ecosystems. For example, coyotes live in a wide range of temperatures and use many food sources. They are fast runners, smart, and social. These characteristics allow them to live in several ecosystems, including forests, grasslands, and deserts.

Saguaro cacti, on the other hand, are specialized for one set of physical conditions. They require hot, dry conditions, with little rain and a bit of shade during their first 10 years. They can survive only in the Sonoran Desert.

One species of organism is unique. *Homo sapiens*, or humans, can live in all the ecoscenario locations. Some environments are unsuitable for humans, so we make better environments for ourselves in these locations. Suitable clothes, housing, and imported food and water allow us to live from the Arctic to the Sahara Desert.

Pine trees harvested for wood and paper products are an example of a provisioning ecosystem service.

Humans and Our Environment

All of us depend on the environment around us. You might think that you do not get much from the environment around you. But think about some basic human needs, such as food and water. For most Americans, drinking water comes out of a faucet. Much of the food we eat comes from a store. But where does the water in your faucet come from? Where does the food in the store come from? They are products of your environment.

For most of us, swimming and hiking are recreational. Recreation is another benefit provided by the environment. Benefits from the environment are **ecosystem services**. Lovely scenery, protection from flooding, and soil formation are also ecosystem services. Ecosystem services include **provisioning**, **regulating**, **cultural**, and **supporting**.

Enjoying the peace and beauty of a wilderness pond is a refreshing and inspiring cultural benefit from nature.

Provisioning services provide supplies that we need, like food from agriculture and wood from logging. *Regulating* services control parts of the ecosystem, like how forests remove air pollution and plants prevent land erosion. *Cultural* services include recreation and inspiration from the environment. *Supporting* services include nutrient cycles like the water cycle and carbon cycle, which are necessary for all other ecosystem services to exist.

Humans benefit from every ecosystem on Earth. Our activities affect the well-being of every ecosystem. The benefits for humans must be measured against the consequences of human actions. Each ecoscenario tells a story about its ecosystem, the services it provides, and how humans affect it.

Think Question

Can you think of other examples of ecosystem services that you get from the environment?

1. Arctic National Wildlife Refuge

The Arctic National Wildlife Refuge is in the northeastern corner of Alaska. It is one of the most undisturbed places on Earth. Its animals live mostly in the 6,070 square-kilometer (km^2) coastal plain.

Biome: Arctic Tundra

Arctic tundra forms a band around the North Pole. It is land that is permanently frozen a few centimeters below the surface. Lichens, sedges (grass-like plants), and low shrubs cover this frozen landscape. Tundra supports large numbers of seasonal insects (mosquitoes, black flies), migratory birds (ducks, geese), resident birds (willow ptarmigans, snowy owls), large mobile mammals (caribou, wolves, brown bears), and smaller mammals (voles, lemmings, snowshoe hares, minks).

Because it is located above the Arctic Circle, the Arctic National Wildlife Refuge has long, cool summer days and long, dark, freezing winter nights. This frigid climate shapes the terrain and organisms found there.

The large snowy owl blends into its frosty tundra surroundings. Because of the open, treeless landscape, the sharp-eyed predator can perch right on the ground to hunt small mammals, mostly lemmings.

The Delaware Water Gap, named for the notch the Delaware River cuts through the mountains, teems with aquatic and terrestrial plants and animals.

2. Delaware Water Gap National Recreation Area

The Delaware Water Gap National Recreation Area is a 64.3-kilometer (km) stretch of the Delaware River. It runs between New Jersey and Pennsylvania and down through Delaware and Maryland to the ocean. The Delaware River is the largest free-flowing river in the eastern United States.

Biome: Fresh Water

The freshwater biome includes rivers, streams, lakes, and ponds. Habitats along the edges of the water are often populated by trees, shrubs, grasses, and vines. The cool, clear, running water contains lots of oxygen for **aquatic** organisms.

Freshwater algae are microscopic. They produce food for many other organisms in the ecosystem. Many **terrestrial** plants feed animals that live on the water's edge.

The Delaware River shore is crowded with trees (hemlock, maple, oak, hickory), shrubs (mountain laurel, rhododendron), and wildflowers.

A rich diversity of animals live in and along the water's edge, including fish (trout, catfish, perch, bass), mollusks (freshwater mussels, aquatic snails), insects (butterflies, mayflies, bees, ants, crickets, cicadas), and amphibians (frogs, toads, salamanders). Many birds (ducks, eagles, swallows, hawks), reptiles (turtles, snakes, lizards), and mammals (squirrels, rats, deer, opossums, beavers, skunks, raccoons) also depend on the river.

In the winter, most of the river freezes over. Summers are warm and humid, with thunderstorms and dense fog. The river flow is generally smooth and quick. Shallow areas and quiet pools are excellent locations for observing wildlife.

3. El Yunque National Forest

Puerto Rico is an island in the Caribbean Sea southeast of Florida. El Yunque National Forest is on the eastern end of Puerto Rico. Warm rain falls all year on the mountains in this area, producing a tropical rain forest.

Biome: Tropical Rain Forest

Tropical rain forest ecosystems have dense vegetation made up of many types of plants (huge trees, vines, shrubs, low plants). Rain forests support large populations of insects (beetles, moths, ants, termites, flies, wasps, butterflies), many types of birds (hummingbirds, orioles, parrots, todies, swifts, warblers, hawks), reptiles (snakes, lizards), amphibians (frogs), and small populations of mammals (bats, rats, cats).

Puerto Rico has a tropical climate. It lies between the Tropic of Cancer and the equator. Rainfall is consistent and heavy throughout the year. Warm temperatures and daylight hours are also fairly constant year-round. El Yunque peak is hidden in clouds. It has slightly cooler temperatures than the forests below.

El Yunque is the only tropical rain forest in our national forest system, and the most biologically diverse. The forest floor trail winds past 240 species of trees; 88 are rare and 23 are found only here.

Alligators are the top predator of the Everglades' freshwater swamps. They also help other populations: their abandoned "gator holes" fill with water, providing shelter and sustenance for other wildlife during dry periods.

4. Everglades National Park

Everglades National Park is a subtropical wilderness on the southern tip of Florida. The Everglades is a slow-moving, shallow, wide river dominated by saw grass. This river starts at Lake Okeechobee and flows south to the Atlantic Ocean, Florida Bay, and the Gulf of Mexico.

Biome: Wetlands

Wetlands are covered in shallow water most of the year. They occur most often in temperate and subtropical regions. They are usually very productive, due to rapid growth of wetland plants. The communities of plants vary throughout the ecosystem, depending on how deep the water is. All of the communities include plants that thrive in or near water (saw grass, mangroves, pine trees).

Wetlands can be densely populated with insects (dragonflies, beetles, mosquitoes, flies), fish (catfish, mosquito fish, bass, perch), reptiles (cottonmouth snakes, turtles, American alligators, lizards), amphibians (frogs, toads), and birds (anhingas, herons, kingfishers, ducks, thrushes, sandpipers). Wetlands are important nurseries for many fish and other aquatic species. Wetlands are also essential feeding grounds for migratory birds.

The Everglades experiences wet and dry seasons. The wet season is warm, and relative humidity can be 90 percent or more.

5. Florida Keys National Marine Sanctuary

The Florida Keys National Marine Sanctuary covers 9,600 km² off the southern tip of Florida. A marine sanctuary is like a national park in the ocean. The reefs in the sanctuary form the third-largest system of coral reefs in the world.

Biome: Coral Reef

Coral reefs develop only in warm, shallow ocean water. Warm seas in the tropics support moderate populations of **phytoplankton**, tiny organisms that carry out **photosynthesis**. They also have large populations of corals. These tiny soft-bodied animals are related to jellyfish. Corals get food from photosynthetic algae, called zooxanthellae. Coral polyps construct protective chambers from calcium carbonate.

These form the hard structure of the reef. The zooxanthellae live inside the coral polyp tissue. In this **symbiotic** relationship, the zooxanthellae produce food for the coral. Billions upon billions of these structures produce huge reefs around the world.

Coral reefs support extremely diverse populations of invertebrates (clams, crabs, shrimp, lobsters, octopuses, snails, sea stars, anemones, urchins, sponges, worms) and fish (rays, groupers, gobies, angelfish, wrasses, eels, surgeonfish, parrot fish, sharks).

The Florida Keys are in a tropical ocean. Temperatures throughout the year range between 27 degrees Celsius (°C) in summer and 20°C–24°C in winter. Sunlight penetrates the water and reflects off the white bottom. This solar **energy** keeps the water warm year-round.

Soft corals, including these sea fans, flutter and sway in the warm ocean waters of the Florida Keys. Like the hard corals that are the building blocks of a reef, soft corals are made up of living polyps.

A deciduous forest biome is always changing because the climate has four seasons with different weather. In autumn the leaves change color, and in winter the trees lose their leaves.

6. Monongahela National Forest

Monongahela National Forest is in the Allegheny Mountains of West Virginia. This national forest covers over 3,678 km². The landscape is rugged, with beautiful views of exposed rocks, spring wildflowers, and colorful fall leaves.

Biome: Deciduous Forest

Deciduous forests, also called hardwood forests, are located in temperate regions. They have moderate rainfall, cold winters, and fairly rich soils. Deciduous trees (maples, oaks, ashes, beeches, birches, hickories, poplars), shrubs, grasses, and low plants dominate this biome. Deciduous forests support large seasonal populations of insects (gnats, black flies, bugs, beetles, moths, butterflies, ants, bees, wasps, termites, crickets) and birds (warblers, thrushes, hawks, owls, sparrows, swallows, swifts, turkeys, grouse, woodpeckers). They support moderate numbers of reptiles (snakes, turtles), amphibians (salamanders, frogs, toads), and mammals (bears, deer, moose, coyotes, foxes, bobcats, skunks, raccoons, woodchucks, squirrels, mice, voles, gophers).

Monongahela has warm summers and cold winters. There is no rainy season. The shape and position of the mountains cause more rain to fall on the western side of the mountains than on the eastern side.

7. Monterey Bay National Marine Sanctuary

Monterey Bay National Marine Sanctuary is off the coast of California. The central coast of California supports a unique ecosystem known as the kelp forest. Kelp grows from the rocky seabed to the ocean surface.

Biome: Kelp Forest

Kelp is a brown algae that carries out photosynthesis. It grows underwater in large groups called forests. Monterey Bay has two dominant species of kelp. The more common species is giant kelp. Giant kelp can grow as much as 0.6 meters (m) a day when it gets enough light. The other species is bull kelp. It grows mostly in exposed areas with lots of water motion from waves.

Kelp forests occur in cold, shallow ocean water. These dense growths of kelp support large populations of many types of invertebrates (sea urchins, lobsters, crabs, octopuses, sea stars, anemones, abalones, snails), fish (sheepshead, garibaldi, sharks, sea bass, rockfish), and marine mammals (seals, sea otters).

The climate in Monterey Bay is relatively mild, with warm summers and cool winters. The waters in the sanctuary are influenced by cold Arctic water and warm tropical water. El Niño can warm the water in the sanctuary by several degrees. El Niño is a weather pattern that occurs every several years.

A world leader in marine research, the Sanctuary is an important economic and recreational resource, too. The kelp forest supports a variety of fisheries, and the annual kelp harvest is used in food and medicine production.

The Sonoran Desert is home to extensive forests of tall saguaro cacti, the classic symbol of the southwestern desert.

8. Sonoran Desert National Monument

The Sonoran Desert is located in Arizona, California, and Mexico. The Sonoran Desert National Monument protects 1,970 km^2 of the desert in Arizona. This hot, dry desert has the most diverse life of the North American deserts.

Biome: Desert

Deserts are rocky, sandy land, with little precipitation and high temperatures. They are populated by large cacti (saguaro, prickly pear, cholla) and tough, drought-resistant trees and shrubs (mesquite, creosote bush). The Sonoran Desert supports a wide variety of insects (ants, beetles, moths), reptiles (rattlesnakes, lizards, desert tortoises), birds (red-tailed hawks, cactus wrens, roadrunners, quail), and mammals (coyotes, bats, jackrabbits, kangaroo rats).

Summers in the Sonoran Desert are hot, with temperatures between 38°C to 46°C. Winters are mild during the day, but can drop to 4°C at night. The rainy season lasts from July to September, but produces only about 25 cm of rain a year.

9. Tallgrass Prairie National Preserve

Covering 45 km² in Kansas, the Tallgrass Prairie National Preserve was created in 1996 as a partnership between the Nature Conservancy and the National Park Service. The grassland is a remnant of the prehistoric prairie that used to cover much of the United States.

Biome: Temperate Grassland

Temperate grasslands, also known as prairies, are great expanses of flat or rolling land covered by fairly rich soil. Grasslands receive moderate amounts of precipitation. Seasonal downpours may be followed by periods of drought. Summers are often hot, and winters cold. Grasslands are dominated by many species of grass and other low plants.

Grasslands support resident populations of small- and medium-sized mammals (ground squirrels, prairie dogs, ferrets, foxes, coyotes) and birds (hawks, prairie chickens, meadowlarks). Their seasonal populations are much larger. They include insects (grasshoppers, bees, beetles, butterflies, moths, locusts, aphids), **migrating** populations of birds (cranes, bluebirds, sparrows, swallows, horned larks), and large, grazing mammals (bison, pronghorn antelope).

The climate of Tallgrass Prairie National Preserve is **semiarid** (*arid* means dry). It has warm summers and cold, dry winters. Prevailing winds during much of the year help make this grassland dry.

Millions of wild bison once roamed the prairies of the American West, but they were overhunted nearly to extinction in the late 1800s. Their populations are now slowly increasing in a few protected preserves.

In some ways, Yellowstone is a typical taiga ecosystem. But the park is most famous for its unique hydrothermal features: geysers, hot springs, and pools brightly colored by microorganisms that live in superhot water.

10. Yellowstone National Park

In 1872, the US Congress established Yellowstone National Park, the first park of its kind in the world. Most of the park is in Wyoming, with small areas in Idaho and Montana. The park includes areas of forest, freshwater lakes and streams, grasslands, and colorful thermal-pond ecosystems.

Biome: Taiga

Taiga is also known as boreal forest. These forests are in northern latitudes, often in mountains. They get moderate rain, and have long, cold winters and short, warm summers. Taiga has dense stands of evergreen trees (pines, firs, spruces). It supports large populations of insects (flies, beetles, moths, wasps, crickets, butterflies), birds (warblers, thrushes, woodpeckers, sparrows, hawks, ravens, swallows), small mammals (squirrels, foxes, weasels, otters, raccoons, rabbits, mice), and large mammals (deer, bears, coyotes, wolves, moose, elk).

Yellowstone snows begin in autumn. By November, deep snow blocks most roads in the park. Snow typically piles up to a depth of 3.8 m over the winter. Snow is also common in the spring. It can fall throughout the summer at the higher elevations.

Defining a Biome

What ecosystem do you live in? You don't have to live in Yellowstone National Park to be in taiga or Sonoran Desert National Park to be in the desert.

The environment where you live has **biotic** and abiotic factors in common with one or more of the ecosystems we will study.

At this point, you have learned the definition of an ecosystem. You know it is a combination of biotic and abiotic factors. Ecologists group similar ecosystems into **biomes**. Biomes have similar climate conditions and communities, although they might not be near each other. Biomes can be terrestrial or aquatic. Biomes are described in terms of abiotic factors like climate and soil type. They differ in biotic factors, like the plant and animal populations that survive in these environments. All the biomes on Earth combine to form the **biosphere**.

Zebras and coarse grasses are populations in the savanna ecosystem that covers half of Africa. This region has much in common with ecosystems in Australia, South America, and India. Together, they make up the tropical grassland biome.

Climate

Weather patterns have been recorded for decades. Measurements include temperature, air pressure, wind, and precipitation. The patterns of weather conditions over long periods of time make up the climate. Climate is one of the most important environmental conditions that define a biome.

Climates vary in part because of their latitude. *Latitude* means how far north or south of the equator a place is on Earth. Earth can be broken into three main climate zones. The **tropical zone** is closest to the equator (latitude 30° south to 30° north). There the Sun shines on land and water about 12 hours a day all year. This constant sunlight creates a stable, warm, wet environment that supports a large **biomass**. Most of Earth's landmasses are found a little farther from the equator, in the **temperate zone** (30°–60° N and 30°–60° S). In this zone, moisture and temperature vary with the seasons of the year. The coldest year-round temperatures are found farthest from the equator, in the **polar zone** (60°–90° N and 60°–90° S).

Abiotic Factors and Aquatic Biomes

Humans are still learning about the underwater biomes that cover three-fourths of the planet. Four ecoscenarios are in aquatic biomes. Abiotic factors that define aquatic biomes include water depth, flow or current, temperature, salinity (concentration of salt), and dissolved nutrients.

Abiotic Factors and Terrestrial Biomes

A terrestrial biome's climate is defined mainly by annual patterns of temperature and precipitation. Precipitation can fall in many forms, including rain, sleet, snow, and hail. It is measured as height of water over an area. In desert biomes, for example, the equivalent of 15 to 25 centimeters (cm) of liquid water falls on the land each year. In contrast, the rain forest biome can receive 125 to 660 cm of liquid water each year.

Temperature is often expressed by an average and a range. Temperature range is the difference between the high and low temperatures. For example, many deserts get extremely hot in the day and very cool at night. A temperature range of 25 degrees Celsius (°C) or more on 1 day is not uncommon. In contrast, most tropical rain forests have a temperature range of 14°C or less all year.

Abiotic factors throughout the tropical rain forest biome are similar. But biotic factors differ from rain forest to rain forest. The sika deer is found almost exclusively in southeastern Asia.

Earth's Biomes

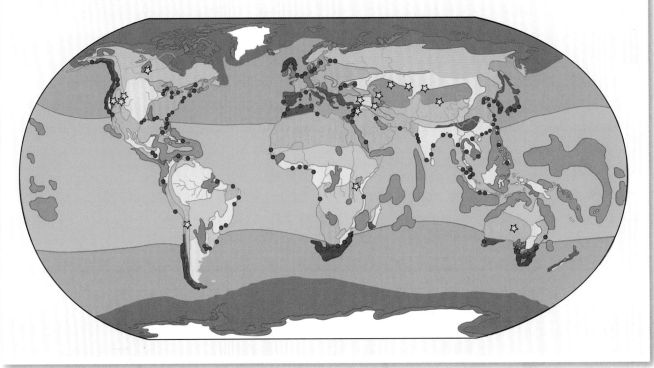

The world map is color-coded to show Earth's biomes. Regions that belong to the same biome may be on opposite sides of the globe.

Climate is rarely the same throughout an entire area. In the Northern Hemisphere, for example, south-facing sides of hills and mountains receive more sunlight than north-facing sides. In a biome like taiga, you might see more trees growing on the moister, shaded north-facing sides of mountains. These small differences are called microclimates. For our study, we will focus only on the larger biomes.

Biotic Factors and Biomes

Biomes often have a particular community of organisms that provides food and shelter for other organisms. Trees dominate forest biomes, kelp dominates kelp forests, and coral dominates the coral reef biome.

Areas that share the same biome share similar characteristics, but they do not have the exact same biotic communities. These variations can be caused by differences in geography, elevation, and local soil.

Think Questions

1. **What biome do you live in?**
2. **How is the biome you live in similar to and different from the nearest ecoscenario location?**

 To learn more about Earth's biomes, including what each color-coded zone on the map above represents, visit FOSSweb.

An Introduction to Mono Lake

Mono Lake is desolate. The desert stretches out from the lake in three directions. The mighty Sierra Nevada mountain range rises up on the fourth side.

The winters are freezing, and the summers are dry and windy. Sagebrush and desert grasses come right down to the salt-crusted shore. The water of Mono Lake is far saltier than the ocean.

Mono Lake is a unique and ancient ecosystem. It is at least 760,000 years old, making it the oldest lake in the United States. The lake is like a puddle left over from a huge, ancient lake complex. It covered a bit of California and a large part of northern Nevada and Utah hundreds of thousands of years ago.

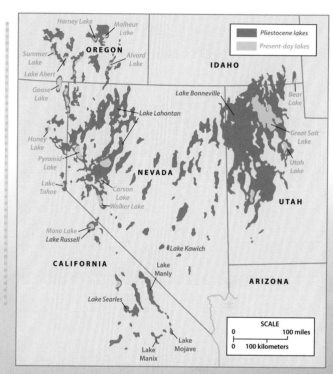

Mono Lake is a small remnant of a massive lake system that once covered the Great Basin Desert.

Throughout its long existence, salts and minerals have washed into the Mono Lake basin from Eastern Sierra streams. Evaporation removes water but leaves deposits behind, so the lake is now almost three times as salty as the ocean.

When the lake complex dried up, only Great Salt Lake, Mono Lake, and a few other small basins retained water. The minerals from the original lake water remain in the modern lakes, but the water amount has decreased. This results in high salt concentrations.

The original lake complex existed for thousands of years. It was a stopping place for huge numbers of migrating birds to feed and refresh themselves before continuing their journey. As time passed, the lake receded, and the land became dry. The migrating birds depended more and more on the remaining pockets of water. Today, these bodies of water are essential to the birds that cross the desert around Mono Lake.

Mono Lake looks barren and empty at first. But it is one of the most productive ecosystems on the continent. Below the water's surface, a huge population of algae is the base of a **food web**. Enormous populations of brine shrimp and brine flies feed on the algae. These crustaceans and insects provide nourishment for the birds that come to the Mono Lake basin. Millions of birds arrive from as far away as the Arctic

A California gull feasts on brine flies, which feast on the lake's algae. The flow of energy through the food web begins with algae.

Circle and the equator. Between midsummer and fall, the birds feast on the abundant food. California gulls and snowy plovers nest on the two Mono Lake islands.

At the lakefront, gray bumpy structures, called **tufa towers**, dominate Mono Lake. The towers are formed over springs on the lake bottom. The calcium-rich water bubbles up through the salt water. The calcium and salt react, making deposits that slowly create tufa towers.

Tufa towers grow only underwater. So why are they sticking up all around the shore of Mono Lake? The answer lies in recent changes in this ecosystem.

Abiotic Factors

Mono Lake is the largest lake (183 square kilometers [km^2]) completely in California. It sits in a broad basin with no outlet, in an active volcanic area. The two islands in the lake, Negit and Paoha, are small volcanic cinder cones. Paoha, the larger island, has steam rising from vents and several hot springs.

Underwater springs rich in calcium mix with lake water rich in carbonates. The chemical reaction produces a kind of limestone called tufa that builds up underwater. The tufa towers were revealed when the lake level dropped.

The most important abiotic factor in Mono Lake is the water's high salt content. The lake contains about 280 million tons of dissolved salts. These salts make the water alkaline. Alkaline water is basic. Basic and acidic substances are described with a numeric scale called **pH**. Pure water is neutral and has a pH of 7. A basic substance you might be familiar with is household bleach, which has a pH of around 12. The water in Mono Lake is not as basic as bleach. The alkali salts raise the pH to about 10. The lake water feels slippery between your fingers.

Rain and snow are critically important to the stability of Mono Lake. Average annual precipitation is around 35 centimeters (cm), which is a very small amount. Fresh water also enters the lake from the many streams flowing from the Sierra Nevada. Spring snowmelt on the mountains is a major source of stream water.

Streams flow into Mono Lake, but no streams flow out. When water in the lake evaporates, the alkali salts remain. Over hundreds of thousands of years, these salts have raised the salinity of the water.

Fresh water coming into the lake offsets the water lost to evaporation. If less water comes into the lake than is lost to evaporation, the lake shrinks. But the amount of salt does not. As a result, the lake water gets saltier.

Life in Mono Lake

No fish or frogs live in Mono Lake. But it is filled with life. The base of the food web is planktonic algae. In winter and early spring, the microscopic algae reproduce quickly. By March, the lake is a thick green soup of single-celled algae.

In March, countless brine shrimp hatch out of cysts on the lake bottom. The microscopic shrimp gobble up the planktonic algae. They mature in a few weeks and produce a second generation. In June and July, 4 to 6 trillion mature brine shrimp, each about 1 cm long, fill the lake.

Meanwhile, the larvae of brine flies (also known as alkali flies) are dormant on the bottom of the lake. In June and July, the larvae start eating algae that grow on the lake bottom. By midsummer, millions of brine flies are skittering across the surface of the lake.

Mono Lake is hemmed in by jagged mountains and harsh desert. Even in winter, life in the ecosystem continues—in the form of a growing algae population.

Eared grebes visit Mono Lake every fall on their migration route to Mexico. They double their weight on the lake's lavish food supply to prepare for the long flight.

Huge flocks of birds arrive to feast on shrimp and flies. Some birds nest at Mono Lake, including 50,000 California gulls (most of California's breeding population and the second largest colony in the world) and 400 snowy plovers. Migratory birds include 800,000 eared grebes (30 percent of the North American population), 80,000 Wilson's phalaropes (10 percent of the world population), 60,000 red-necked phalaropes (3 percent of the world population), and smaller numbers of 79 other species of waterbirds.

Between April and October, the bird activity is constant. When the last migrating birds leave at the end of October, Mono Lake is quiet for a brief time. The surface is lifeless—no flies and no birds. The lake water is clear and still. The brine shrimp ate almost all of the algae. The migrating birds ate most of the brine shrimp. Brine shrimp egg cysts and dormant brine flies wait on the lake bottom. Autumn closes in, and Mono Lake gets cold. The cold surface water sinks to the bottom. It pushes up nutrient-rich water from the lake bottom. Nutrients rise to the surface for the 760,000th time, and the few remaining algae start to reproduce. The cycle repeats.

Think Questions

1. **What would happen if the amount of fresh water entering Mono Lake increased?**
2. **What would happen if the amount of fresh water entering Mono Lake decreased?**

Biosphere 2: An Experiment in Isolation

Did you interact with Earth on your way to school this morning? Did you walk on the ground? Did you breathe the air? Was it raining?

As a living organism, you are always interacting with Earth and Earth's systems. Earth is a set of interacting subsystems. The **geosphere** is the rocky, mineral subsystem of the planet. This is the hard part of the planet that we live on. Earthquakes and volcanoes remind us that the geosphere is always changing.

Wrapped around Earth is the **atmosphere**. This thin layer of gases extends about 600 kilometers (km) above Earth's surface. The atmosphere provides essential gases. It is an energy-transfer system and an insulator. It shields us from extraterrestrial radiation. It is also an important medium for water distribution.

Earth is a water planet. On Earth, water exists naturally in three states: liquid, gas, and solid. All the water on Earth makes up the **hydrosphere**. The hydrosphere includes the ocean, lakes, rivers, streams, and water underground. It includes polar icecaps, glaciers, snowpacks, and permafrost. It also includes water vapor in the air and water drops in clouds, fog, and precipitation.

Earth's vast ocean is the chief component of its hydrosphere. Waves are an example of system interactions, in this case between the ocean surface and wind in the atmosphere.

And finally, creeping, running, burrowing, flying, and swimming in the other three spheres is the biosphere. It includes all the living organisms on Earth. Millions of different life-forms give Earth its particular flavor.

All four subsystems are bundled into one global system. The biosphere interacts with the other three subsystems. Life is always connected to the physical environment.

Biosphere 2

To study Earth's biosphere, a group of scientists decided to study an entire ecosystem. You can do an **observational study** or **population study** in the wild. But these scientists wanted to set up a **controlled experiment** on an ecosystem.

A real ecosystem is much larger and more complex than a classroom habitat. Say you want to find out how carbon dioxide (CO_2) affects an ecosystem. You could add CO_2 to a field or forest to observe changes in plant growth. But the wind would soon blow the CO_2 away. You could monitor the weather for 40 or 50 years and look for patterns. But it would be impossible to control variables.

Biosphere 2 in Arizona is a unique research facility containing seven model land and water ecosystems.

Say you noticed that plant growth and health in a cold, rainy year differs from a hot, dry year. Is the difference because of rain, moisture in the air, temperature difference, or amount of sunlight? It is impossible to say, with so many variables changing at once.

A controlled experiment can help you determine cause and effect. To do controlled experiments, you need a terrarium that is big enough to hold a whole ecosystem. Then you could control all but one of the variables. Biosphere 2 is that terrarium.

Biosphere 2 was built to answer some big questions. Can human beings and other organisms live in an airtight building for 2 years with no outside support? Can they set up an ecosystem that provides for their every need and the needs of the other organisms around them?

In 1984, construction began on Biosphere 2, the world's largest **model** of Earth as a system. Scientists use models to develop explanations about the natural world. Models make it possible to explore complex aspects of the real world that are not easily observed.

Design Challenges

To create an accurate model, the Biosphere designers faced several major challenges. The first challenge was to choose a site and design a structure that resembled Earth. It would be an open system for energy and a closed system for matter. Energy could enter and leave, but no materials (even gases) could pass through. The chosen site was in the Sonoran Desert, in Arizona, where the sky is often cloudless. This site maximizes access to sunlight. Sunshine is essential to **sustain** the humans and hundreds of other species in the closed system. It provides essential energy.

Sealing the giant terrarium against the outside environment was also essential. Biosphere 2 is a greenhouse built on a solid steel foundation. It is completely isolated from its surroundings.

The second design challenge was to create a self-sustaining ecosystem that could support four men and four women for 2 years. Ecologists carefully selected the organisms for Biosphere 2. They needed plants, animals, and microorganisms. They had to provide food for the human team, and supporting services for the organisms that would produce that food. They needed organisms to refresh the air and dispose of waste materials. The planning was complex and detailed. Many lives depended on getting it right.

Biosphere 2 covers almost as much area as four football fields. Inside are seven environments: rain forest, desert, tropical ocean, marsh, savanna, thorn scrub, and a farm within a forest. The farm area was planted 1 year before the mission began, so that food would be available from day 1.

Biosphere 2 has the world's largest greenhouse, with 6,500 windows! Inside the huge structure is equipment that controls environmental factors in all the enclosed ecosystems.

The Biosphere 2 campus is located about an hour from Tucson, Arizona, and now offers tours and classes to the public.

The First Mission

In September 1991, the door was closed. The eight humans and 1,800 other populations were sealed on the inside. In many ways, these humans became Earth astronauts. The challenges they faced were like those in long-term space travel. The trip to the Moon takes a few days. A trip to Mars might take a year. The most efficient way to make such a journey would be in a tiny ecosystem that recycles everything needed for life.

Low oxygen levels. After only 1 day, the team noticed that CO_2 levels were rising quickly and oxygen (O_2) levels were dropping. If they could not solve this problem, the entire project would fail. Where was the O_2 going?

Part of the answer was in the soil. The populations of soil microbes were growing too fast, using too much O_2. The scientists reasoned that if the O_2 concentration was going down, the CO_2 concentration should be going up. But it was not going up as fast as expected. Scientists later discovered that the building's concrete was absorbing CO_2. Concrete cures slowly, a hardening process that uses CO_2.

Take Note

Compare your classroom minihabitats with Biosphere 2. What similarities do you see? How do the models differ?

Knowing the source of the problem did not solve it. At first, O_2 concentration was 21 percent, the normal amount at sea level. After 16 months, it had fallen to 14 percent, the amount found at 5,000 meters (m) elevation. CO_2 levels were 12 times higher than those in Earth's atmosphere. In the end, O_2 was brought in from the outside world. This change in the experimental setup allowed the humans to continue under safer conditions.

Uninvited guests. Two uninvited local species, cockroaches and longhorn crazy ants, got into Biosphere 2 and caused problems. The cockroach population grew out of control. Crazy ants are excellent communicators, which gave them an advantage over the other insect species in Biosphere 2. The crazy ant population increased while most other arthropod populations decreased. The ants also clogged vents and chewed on wiring.

Small successes. The Biosphere team made it through the 2-year mission. But they got some outside assistance and a lot of criticism. Still, a few great insights came from this first large-scale model of Earth. While the team struggled to produce food the first year, they produced more than 80 percent of their food the second year. Their low-calorie,

nutrient-rich diet significantly lowered the cholesterol and blood pressure of several people, and strengthened many of their immune systems.

In addition, the airtight Biosphere 2 allowed scientists to measure O_2 and CO_2 concentrations throughout a day and over a season. These changes in atmosphere were connected to the interactions of plants, animals, and soil microorganisms.

Later development. A second mission in 1994 lasted 6 months. After that, Biosphere 2 was no longer a live-in facility. It was transformed into a center of research on Earth's living systems. Today, Biosphere 2 is the world's largest Earth science laboratory.

Recent Research on Carbon Dioxide

You have probably heard about humans using energy sources such as gasoline and coal that release CO_2 into the air. High CO_2 levels in the atmosphere are changing the global climate. What will happen if CO_2 levels continues to rise? Scientists hope to use Biosphere 2 to gather information to help answer this question.

Cockroach invasion and then overpopulation were unexpected problems scientists had to solve in the early phases of Biosphere 2.

Tropical rain forests. Tropical rain forests have a big effect on the atmosphere. They take in a lot of CO_2 and produce a lot of O_2 during photosynthesis. Dense vegetation and long days of direct sunlight year-round are typical in the tropics. Plants in the rain forest carry out photosynthesis at a higher rate than anywhere else on Earth. Some scientists think rain forests have great potential for controlling rising CO_2 levels.

How will changing CO_2 levels interact with changing climate patterns? In the Biosphere 2 rain forest, scientists are researching how changes in rainfall affect photosynthesis of rain forest plants. They can control the amount of "rain" with the sprinkler system. They can pump in more CO_2, and regulate the temperature. Scientists change these variables one at a time. They look at the effect each variable has on the plants and the amount of CO_2 in the atmosphere.

Coral reefs. Coral reefs have been called the rain forests of the sea because they support so much **biodiversity**. Coral defines the reef ecosystem. Algae, fish, crabs, and many other sea organisms live on, in, and around the coral structures.

The Biosphere 2 rain forest enclosure allows scientists to experiment with conditions likely to become more common in the world's tropical rain forests.

The Biosphere 2 ocean was designed to simulate a Caribbean coral reef environment. Here, marine biologists are exploring ways to reverse the global decline of reef habitats.

Marine biologists report that coral growth has slowed around the world. Fewer and fewer fish and other organisms are living in coral-reef ecosystems. Warming oceanic waters have been linked with coral death. Could more CO_2 in the water be affecting coral-reef ecosystems? Several scientists are studying the coral reefs in the ocean biome at Biosphere 2. It holds 2,500,000 liters (L) of water. The depth ranges from 0 m at the beach to 7 m. Scientists have investigated several factors that might affect the health and survival of the coral.

When the concentration of CO_2 in the water increased, the health of the coral declined. Excess CO_2 prevented coral from getting calcium to build its skeleton. Studies of the coral reefs in Biosphere 2 provide evidence that changes in the ocean, such as more dissolved CO_2, are damaging coral-reef ecosystems.

Important conclusions. As scientists learn more about ecosystems, two things become very clear. Any change in one part of an ecosystem affects every other part of the ecosystem. These changes are often impossible to predict. The more we learn, the more we realize how complex natural ecosystems are. We still understand very little about the way ecosystems work or how human activity affects them.

Why should we care? What difference does it make if rain forests or coral-reef organisms are disappearing? That's not where we live.

Well, it is where we live. Earth is small. The atmosphere surrounds the planet. The ocean washes the shores of all the continents. Changes in one ecosystem are communicated to the rest of the world by flowing air and water. Everything is connected. Small changes in global temperature can have a huge effect on weather patterns. Weather distributes water, and water is life.

Maybe you can join the community of people working on tough ecosystem problems. College students attend classes at Biosphere 2 for a semester to study environmental problems. Biosphere 2 also has a summer program for high school students who are interested in studying the environment.

Think Questions

1. **What are some benefits of doing research on ecosystems in Biosphere 2 rather than in a natural ecosystem?**
2. **What are some limitations of doing research on ecosystems in Biosphere 2 rather than in a natural ecosystem?**
3. **Why is the increase of CO_2 in Earth's atmosphere such an important problem?**

A Rough Comparison of Earth and the Biosphere 2 Original Experiment

System Variables	Earth System	Biosphere 2
Location	Earth, solar system	Oracle, Arizona
Footprint	126 billion acres	3.1 acres
Biomes	21	5
Species	Approx. 10–40 million	1,800
Human population	Over 7 billion	8
Energy sources	Sun, fossil fuels	Sun, fossil fuels
Age of life	3.5 billion years	2 years
Enclosure	Magnetic field, gravity	Glass and steel frame, gravity

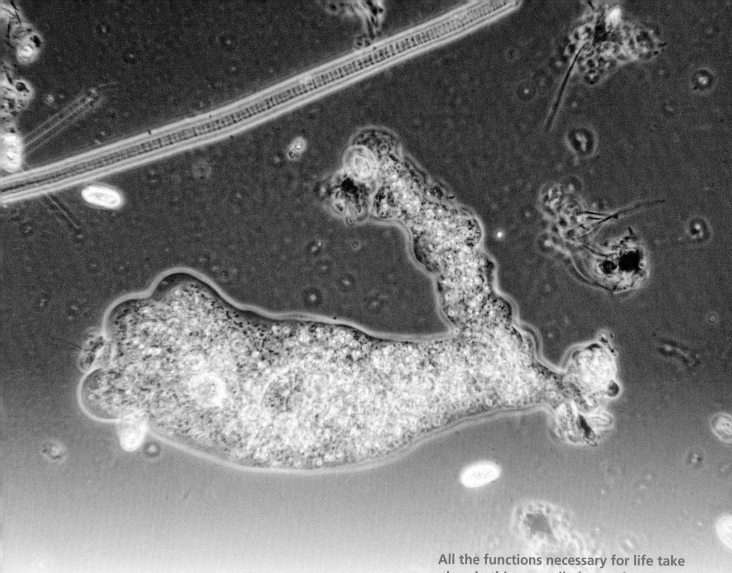

Energy and Life

We all need energy. Even the single-celled amoeba needs energy. But this amoeba isn't going to eat a cheeseburger and fries for dinner.

So, how does it get food? The amoeba eats by encircling bacteria or other **protists** and absorbing the nutrition.

Energy is the ability to do work. Organisms need energy to perform the basic functions of life. These processes include growing, reproducing, exchanging gases, eliminating waste, getting water and nutrients, and responding to the environment. An amoeba manages all this in a single cell. All living cells require energy, because they are all constantly working.

Food is the source of matter and energy for all organisms. But not all organisms eat food. Some organisms make food right in their own cells.

Producers

The Sun is the source of energy for most organisms in an ecosystem. Some organisms capture energy from the Sun, using photosynthesis (*photo* = light; *synthesis* = putting things together). Plants, algae, and phytoplankton typically make their own food this way. These organisms are **producers**.

Most producers transfer energy from sunlight into chemical bonds of food **molecules** (**carbohydrates**). Photosynthesis creates these new molecules. It uses sunlight and **chlorophyll** to turn carbon dioxide (CO_2) and water (H_2O) into carbohydrates. Here is a simplified equation for this chemical reaction.

$$6CO_2 + 6H_2O + Light \longrightarrow C_6H_{12}O_6 + 6O_2$$

carbon dioxide   *water*   *energy*   *sugar*   *oxygen*

Photosynthesis takes place in the green parts of a plant, mostly the leaves. In a cactus, this food production process happens in the trunk or stem.

Water molecules and carbon dioxide molecules change into a carbohydrate molecule (sugar) with energy from light. The process also produces molecules of oxygen (O_2).

Many sugar molecules are used for food. The food has potential energy in its chemical bonds. Photosynthetic organisms use their sugar as food to get energy and the building blocks they need. Plants never eat sandwiches, and photosynthetic bacteria never eat fruit. They don't have to. Because they make their own food, producers are also known as **autotrophs** (*auto* = self; *troph* = food).

Producers create biomass. Biomass is the matter produced by organisms in an ecosystem. When producers build a new carbohydrate molecule through photosynthesis, their biomass increases. Producers transform the Sun's energy into potential energy in biomass.

Producers on land include plants such as grasses, shrubs, and trees. Producers that live in water include phytoplankton, aquatic plants, sea grass, and seaweed.

Consumers

Humans are not autotrophs. We cannot make food in our own bodies. The same is true for all animals, all fungi, and most protists and bacteria. Organisms that do not make their own food eat the organisms that do. Organisms that eat other organisms are **heterotrophs** (*hetero* = other; *troph* = food). Another name for heterotrophs is **consumers**.

Aerobic Cellular Respiration

Producers make life possible for consumers through photosynthesis. But producers do not use photosynthesis as a favor to us. They need the food energy as well. Once the energy is stored in the bonds of molecules, what do producers do with it?

Producers and nearly every other organism on Earth use the process of **aerobic cellular respiration**. During this process, organisms release the stored energy of carbohydrates. The bonds of the carbohydrate molecule break, and new products form (CO_2 and H_2O). Aerobic cellular respiration releases energy. Cells use this energy to do their work. In the process, energy is lost to the environment as **thermal energy** (heating).

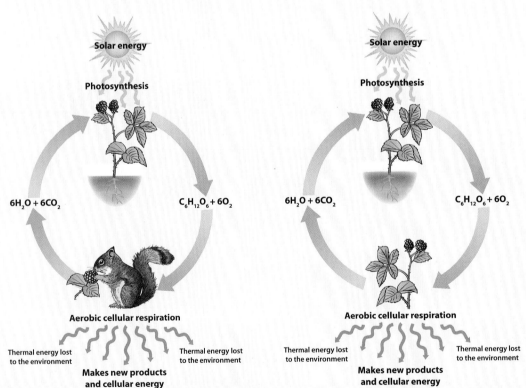

Obtaining Usable Energy

Solar energy

Photosynthesis

$6H_2O + 6CO_2$ $C_6H_{12}O_6 + 6O_2$

Aerobic cellular respiration

Thermal energy lost to the environment

Thermal energy lost to the environment

Makes new products and cellular energy

Solar energy

Photosynthesis

$6H_2O + 6CO_2$ $C_6H_{12}O_6 + 6O_2$

Aerobic cellular respiration

Thermal energy lost to the environment

Thermal energy lost to the environment

Makes new products and cellular energy

Both the squirrel and the berry plant get energy for life through aerobic cellular respiration: using oxygen to break down sugars. The difference? The plant produces and stores the sugars it uses for food; the squirrel must eat the plant to get sugars.

Here is a simplified equation showing what happens in the cell during aerobic cellular respiration.

$$C_6H_{12}O_6 + 6O_2 \longrightarrow 6CO_2 + 6H_2O + \text{usable energy}$$

sugar   oxygen   carbon   water   energy
dioxide

Almost all organisms rely on aerobic cellular respiration, which requires sugar. But only photosynthetic organisms can capture the Sun's energy to create sugars. How do other organisms get sugars? All other organisms must eat photosynthetic organisms, or eat other organisms that already ate photosynthetic organisms.

Think Questions

1. **Look closely at the equations for photosynthesis and aerobic cellular respiration. What do you notice?**

2. **Why do autotrophs, like plants, make sugars?**

Trees take in carbon dioxide and release oxygen into the air during photosynthesis. So healthy, mature forests play an important role in maintaining Earth's atmospheric oxygen.

Depending on the fillings you choose, a taco can be a well-balanced and nutritious meal—a delicious source of the matter and energy we get from food.

Where Does Food Come From?

We love to eat, and we eat to live. Every animal has to eat to live. We may eat different things. Your favorite food might be tacos, but your friend's might be pizza.

But what we eat has one thing in common. It's all food and gives us energy.

Food is the source of two essential elements of life: molecular building blocks and energy. Molecular building blocks are matter. Like all matter, they are made of **atoms**. Building blocks are molecules that come from the nutrients in food. Without these molecular building blocks, you could not grow or replace cells. Your body could not regulate body processes or heal when you get injured.

Energy is not matter. It is not made of atoms. Energy cannot be created or destroyed, but it can be transferred. Light energy from the Sun enters an ecosystem. Organisms convert it to energy stored in the bonds of food molecules. All the things that living organisms do require energy. It is needed to move, talk, digest food, keep warm, think, grow, and feel. All organisms return energy into the environment as thermal energy (heating). Energy is the driver that keeps the machinery of life running.

Pizza dough is typically made from wheat flour, which is finely ground wheat seeds. Breads, pastas, cereals, tortillas, and other foods from similar plant sources are called grains.

Following the Energy Path

Much of Earth's energy comes from the Sun. But humans get their energy from food, not directly from the Sun. How did the energy get into our food? Most people in the United States get food at grocery stores and restaurants, but where does that food come from? What organisms stored light energy in the bonds of food molecules that we eat? Let's find out by following the paths that lead to a pepperoni pizza.

The main ingredients in a pepperoni pizza are bread dough, tomato sauce, cheese, and pepperoni sausage. Start with the dough . . . That's a short path. The pizza dough is bread, made from wheat flour. Wheat flour is made by milling or grinding the seeds of the wheat plant, a kind of grass. So the crust of the pizza comes fairly directly from a plant source.

Tomato sauce is made by grinding tomatoes into a pulp and cooking them with seasonings. The tomatoes are the fruit of a tomato plant. The seasonings, such as oregano and basil, are leaves of other plants. So the sauce of the pizza also comes fairly directly from a plant source. The tomato plant is like all other photosynthetic organisms. It uses sunlight, water, and carbon dioxide (CO_2) to store energy and create the building blocks it needs to grow. When we eat fruit, those building blocks and stored energy of the plant become our food.

Cheese is made of milk, mostly from cows but also from sheep, goats, and other grass-eating mammals. Where do you suppose buffalo mozzarella comes from?

Cheese is processed from milk, so the nutrients in cheese come from milk. Milk is produced by cows that eat grass. The nutrients in milk come from that grass. So, we can trace the energy and nutrients in cheese back to the first organism in the line, a plant.

And what about the pepperoni? Pepperoni is made from the ground-up muscle of cattle and pigs. The animals grew by eating grasses and grain seeds such as corn and millet, which are plant sources.

It looks like each ingredient in a pepperoni pizza has its origin in a plant source. Plants capture the energy in sunlight and transfer it into the bonds of food molecules. This process is photosynthesis. Plant cells use sunlight to rearrange water and CO_2 molecules into molecules of sugar (food),

releasing oxygen (O_2). The sugar molecules are building blocks that support plant growth and store energy.

When an animal eats a plant, those molecules and stored energy transfer to the animal. If another animal eats that animal, the molecules and stored energy transfer yet again. So when you trace any food back to its source, it always starts with plants or plantlike organisms.

Did you order anchovies for your pizza? These little fish do not eat like cows and pigs. Anchovies are animals that eat other animals and plants by filtering the food from ocean water. The animals they eat are tiny free-swimming critters called **zooplankton**. The plants they eat are also tiny free-floating organisms called phytoplankton.

And what do the zooplankton eat? They also eat phytoplankton. These plantlike organisms make their own food by photosynthesis. The anchovies arrive on your pizza because of the food in the microscopic phytoplankton floating in the sea, providing matter and energy.

Eating for Energy and Nutrients

Plants and other photosynthetic organisms use the energy in sunlight to turn water and CO_2 into food in the form of sugar. Some of the energy is stored in the bonds of the sugar molecules.

Carbohydrates. Organisms that cannot produce food must consume other organisms. At this simple level, it might seem that the whole system is based on sugar. But there is more to the story. Sugars are simple carbohydrates. These sugars can be used as building blocks to create other, larger molecules that the plant needs, including starches and cellulose (plant fiber). These larger molecules are called complex carbohydrates. Most plant carbohydrates are so complex that human stomachs cannot digest them. This is one reason we cannot just go out on the lawn and eat grass for lunch.

Complex carbohydrates like fruits and vegetables keep the body fueled for extended periods. Their high fiber content helps with digestion and keeping the heart healthy.

These sources of protein help the body grow and repair cells.

Simple sugars are the fastest source of energy, but they do not last long. When a person eats simple sugars, cells can access the energy immediately. They also burn up the energy quickly. So you might feel a quick burst of energy after consuming candy or a soft drink, but run out of energy soon after. If the body consumes more simple sugar than it can process, it will store the excess as fat. Complex carbohydrates provide a steady amount of energy to the body throughout the day as they are digested. Good sources of complex carbohydrates are vegetables, fruits, and whole grains such as brown rice, oatmeal, and whole-wheat bread.

Proteins and fats. Carbohydrates are the foundation of energy stored in food, but your body needs other substances, too. Proteins and fats are other molecules produced by organisms. Eating these organisms transfers their proteins and fats.

Protein is important for repairing and building cells. People can get plenty of protein from regular healthy meals. Good sources of protein are eggs, dairy products, beans, nuts, soy, and meat (fish, beef, pork, and poultry).

Fats provide slow, long-lasting sources of energy. You also need fat in order to digest proteins. Everyone needs a certain amount of fat each day. Some fats, including fish oils and nuts, are healthier than others.

Other nutrients. Vitamins and minerals are in the food you eat. These nutrients help your cells get energy from the food you eat, and keep you from getting sick. You do not get energy from these nutrients. But if you are not getting enough of them, your cells cannot effectively use energy. This can make you feel weak or tired.

Two important nutrients for active people are calcium and iron. Calcium helps build bones, and iron carries oxygen to muscles. Many teens do not get enough of these minerals. They are not found in snack foods like soft drinks, potato chips, and candy. Excellent sources of the vitamins and minerals you need include a variety of fresh foods, especially vegetables and fruits. Leafy green vegetables, such as spinach or kale, are a great source of many minerals and vitamins, including iron and calcium. You can also get iron from meat. Calcium can be found in some fruits, such as blueberries. Good sources are dairy foods like milk, yogurt, and cheese.

Ahh, thinking about nutrients and energy made me hungry. Maybe it's time for a healthy snack!

Think Questions

1. **How does the energy stored in food enter the ecosystem?**

2. **What was the last food you ate? Draw a diagram to show a possible pathway of energy from when it entered the ecosystem to when you ate it.**

Milk is a good source of calcium, a mineral the body needs to build and maintain strong bones and teeth.

What Does Water Do?

It's a hot spring day, and you just finished gym class. As you head indoors, you and your classmates swarm to the water fountain for a sip of cold water.

Just a few sips make you feel a bit cooler, less tired, and more comfortable. Why does water make a person feel so much better? Why does thirst make a person feel so terrible? Only food can give your body energy, but your body also needs water. Your blood, which contains a lot of water, carries oxygen to all the cells of your body. You need water to digest your food and get rid of waste. Water is the main ingredient in the sweat that helps cool your body on hot days. Each cell in your body depends on water to function normally.

Life as we know it would not exist without water. More than half of your body weight is water. You could not survive for more than a few days without consuming water! Any fluid you drink contains water, but plain water or milk are the best choices. Many foods contain water, including fruits and vegetables.

On an average day, about one-quarter of our water intake comes from food. Fruits like strawberries are mostly water.

How Much Is Enough?

Since water is so important, you might wonder whether you are getting enough. How much fluid you need depends on your age, size, level of physical activity, and environment. For example, when you heat up by exercising, your body will sweat more to help cool you down. Sweat is a form of water loss from your body. You can lose a lot of water in just an hour of activity on a hot day. If you do not replace that water, you can become dehydrated.

Dehydration means the body does not have enough water to function normally. The body is used to small changes in water content. But when the water loss becomes too great, dehydration can make you feel sick.

Besides sweating, illness is another common cause of dehydration in teens. Sick people can lose fluid through vomiting and diarrhea. You might not want to eat or drink if your throat hurts or if you keep vomiting. But remember that many ailments improve when you drink water.

Drinking water while exercising keeps your body hydrated.

A common symptom of dehydration is a headache. Other symptoms include feeling dizzy, having a dry mouth, and producing less and darker urine. As dehydration gets worse, a person will start to feel more and more sick. Eventually, most organs and body systems will be affected by the lack of water. In severe cases, a person may need to recover in a hospital.

Preventing Dehydration

The easiest way to avoid dehydration is to drink water often, especially when you get thirsty on hot days. When you drink is also important. If you are going to sports practice or working hard outdoors, drink water before, during, and after. It is much better to drink small amounts often than to down a lot at once.

Most of the time, your body keeps you properly hydrated without you knowing it. The body holds on to water when you do not have enough and gets rid of it if you have too much. The water you need comes mostly from food. The rest comes from drinking a glass of water or other liquid whenever you get thirsty.

Other Drinks

People are often tempted by drinks that promise an energy boost or quicker rehydration. Most of these drinks contain sugar. When you are quite thirsty and gulp down that drink, you may be getting a lot more **calories** than you realize.

Energy drinks and soft drinks often contain caffeine, as well as calories. Caffeine is a substance that is produced in the leaves and seeds of many plants. Tea, coffee, cacao (used to make chocolate), and kola (a tree whose fruit gives cola its name) produce caffeine. Chemists also produce caffeine artificially, and it can be added to food or drinks.

Caffeine acts as a stimulant. It can make you feel more awake and alert. But it does not contain energy the way food does. People who consume large amounts of caffeine may feel jittery and have trouble sleeping.

When you feel a little run down, try drinking some water before reaching for an energy drink. Check the nutrition label before you consume a drink to make sure it does not have a lot of sugar or caffeine.

Think Questions

1. **Review your observations of the classroom habitats. What evidence do you have that all living organisms need water?**

2. **Why do you feel like you have more energy after drinking water?**

Caffeine, found naturally in cacao (cocoa) beans and other plants, is the stimulant in coffee, tea, soft drinks, energy drinks, and chocolate. Caffeinated drinks can cause headaches and insomnia in some people.

Wangari Maathai: Being a Hummingbird

Have you heard about Professor Wangari Maathai (1940–2011)? She won a Nobel Peace Prize and has inspired thousands of people around the world.

How? She saw a problem in her local ecosystem and decided she might be able to fix it. Maathai grew up in Kenya, a country on the east coast of Africa. By the late 1970s, most of the trees in the forests of Kenya and surrounding areas had been cut down for firewood and buildings. Without trees, rain washed away the rich soil. Other plants struggled to survive. The loss of soil and plants also meant that local streams began to dry up. Wildlife disappeared, and it became hard to find or grow food.

Wangari Maathai's Green Belt Movement empowers communities, particularly women, to conserve the environment and improve their livelihoods.

In her persistent devotion to environmental activism and human rights, Maathai was inspired by the tiny but tireless hummingbird.

Maathai noticed that this situation was hardest for women. They were responsible for seeking water, firewood, and food for their families. It was an overwhelming situation, but Maathai was inspired by a story. Here is the story in her own words.

The story of being a hummingbird is about this huge forest being consumed by a fire. All the animals in the forest come out, and they are transfixed as they watch the forest burning. And they feel very overwhelmed and powerless, except for this little hummingbird. It says, "I am going to do something about the fire." So it flies to the nearest stream, takes a drop of water, and puts it on the fire. It goes up and down, up and down, as fast as it can. In the meantime, all the other animals—much bigger animals, like the elephants, with a big trunk that could bring much more water—they are standing there helpless. And they are saying to the hummingbird, "What do you think you can do? You are too little! This fire is too big! Your wings are too little, and your beak is so small, you only can bring a small drop of water at a time!" But as they continue to discourage it, it turns to them without wasting any time, and tells them, "I am doing the best I can."

Maathai decided to be the hummingbird and do the best she could to help her fellow Kenyans. She knew how important a strong community of producers, such as trees, is for a healthy ecosystem. So she started a small organization that taught women about basic ecology and how to plant and care for trees.

The organization, called the Green Belt Movement, was a great success. But life was hard for Maathai. She was the first woman in East and Central Africa to earn a doctorate degree. It was also rare for a woman to be so outspoken. Her husband divorced her, and she faced public criticism. She was arrested and beaten several times by the police for her political activism. Still, she continued to work tirelessly toward her goal, just like the hummingbird in her story. She became known as Mama Miti, or "Mother of Trees."

Today, the Green Belt Movement has planted more than 30 million trees in Kenya and other African countries. The local ecosystems are becoming healthy forests again. As the movement grew, it helped educate and empower women across Africa in many ways. It encouraged them to understand their local environmental problems, take action to seek solutions, and become leaders in their communities.

Maathai recognized the importance of healthy forests. Her movement's tree-planting activities in critical watersheds of Kenya are transforming the African landscape.

In 2004, Maathai became the first African woman and environmentalist to win the Nobel Peace Prize. In her acceptance speech, she said, "I would like to call on young people to commit themselves to activities that contribute toward achieving their long-term dreams. You are a gift to your communities and indeed the world. You are the hope and our future."

Think Questions

1. Can one hummingbird make a difference? Explain your thinking.

2. How could protecting the producers of an ecosystem affect the entire ecosystem? Use specific examples from the article.

3. The Nobel Prize committee explained that Wangari Maathai "thinks globally and acts locally." Can you give an example of how you could think globally and act locally?

Rachel Carson and the Silent Spring

Spring is a time of renewal in an ecosystem. Leaves grow, flowers bloom, birds begin to build nests, and hibernating animals emerge from their dens.

Spring can be a noisy season with all this activity. What if this cycle of life was disrupted?

During World War II, the US government began making chemical pesticides. The pesticides were used to kill insects that could transmit diseases to people, including soldiers. After the war, farmers used these pesticides to protect their crops. They killed insects, weeds, and small animals that damage crops. One widely used pesticide was DDT.

Rachel Carson (1907–1964) was a biologist and author. She became concerned about pesticides' effects on the environment. Huge amounts of DDT were being sprayed on crops and playgrounds during mosquito season. Carson wondered how DDT was affecting people and the environment. She decided to find out.

Carson found that excess DDT entered the ground and water system. From there, earthworms and aquatic insects brought it into the **food chain**. Birds and fish ate the contaminated worms and insects. They accumulated more and more DDT in their bodies. Birds of **prey** that ate those birds and fish accumulated even more DDT. This is called **biomagnification**. Organisms at the top of the food chain include humans and bears. These organisms accumulate the highest concentrations of DDT and other toxins.

During the 1950s, many bird populations sharply decreased. High levels of DDT in the birds' bodies affected their eggs. The eggs were so fragile that they broke in the nest before hatching. Many people were especially concerned about the bald eagle population. The bald eagle is a US national symbol. By 1963, habitat destruction, illegal hunting, and the effects of DDT had reduced bald eagle populations. Just 147 breeding pairs were left in the United States (not including Alaska).

Pesticides are widely used in agriculture to protect crops. But many also have been linked to human health hazards and harmful effects on soil, water, and other organisms.

Biomagnification

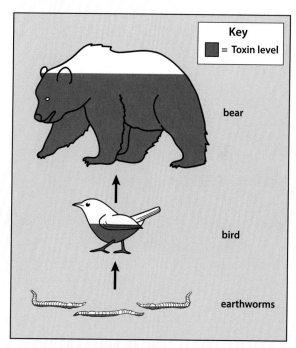

Concentrations of toxins in consumers increase as we move up a food chain.

Anything but Silence

Carson spent 4 years gathering data. She wrote *Silent Spring* about her research. The book focused on environmental pollution with DDT and other pesticides. She warned that we might someday have springs without birds singing. The book came out in 1962 and quickly created a lot of public debate.

Chemical manufacturers were furious. They ran ads telling Americans to ignore Carson's work. They tried to discredit her ability as a scientist. They assured the public that pesticides were perfectly safe.

But Americans found the evidence quite convincing. The White House and Congress were flooded with letters from anxious citizens demanding action. President John F. Kennedy (1917–1963) appointed a committee of scientists to investigate Carson's claims. Congress also formed an investigative committee.

Carson spent the rest of her life defending her findings. When she testified before the US Senate, she declared, "I deeply believe that we in this generation must come to terms with nature." In defending her research, Carson told Americans to think for themselves. "As you listen to the present controversy about pesticides," said Carson, "I recommend that you ask yourself: Who speaks? And why?"

The recovery of the bald eagle is an American success story. Thanks to protections under the Endangered Species Act, our national bird has come back from the brink of extinction and is flourishing.

Responding to New Awareness

Carson's ideas may not seem surprising today. But before 1962, few people had thought much about pollution and other environmental problems. US industries were constantly introducing useful and exciting products. Few people stopped to think about negative side effects of the products.

President Kennedy's committee found evidence that supported Carson's warnings. So did other government studies. Congress began passing laws to ban or control the use of potentially harmful pesticides. In 1967, bald eagles became protected under the Endangered Species Act. In 1970, Congress established the Environmental Protection Agency (EPA). The agency works to reduce and control pollution of water, air, and soil. In 1973, DDT was banned in the United States.

What happened to the eagles? The ban on DDT helped bald eagle populations recover. Programs of captive breeding and habitat protection also helped. Today, there are about 10,000 wild breeding pairs south of Canada. The bald eagle has been removed from the threatened and endangered species list.

Rachel Carson's *Silent Spring* warned of the dangers of chemical pesticides and launched the environmental movement.

After her death, Carson was awarded the Presidential Medal of Freedom by President Jimmy Carter (1924–). President Carter described Carson as "a biologist with a gentle, clear voice, she welcomed her audiences to her love of the sea, while with an equally clear determined voice she warned Americans of the dangers human beings themselves pose for their own environment."

Many years have passed since *Silent Spring* was published. People are more aware of our effects on the environment. But we continue to invent new chemicals with unknown side effects. And environmental regulations can be removed by the federal government. Can we avoid the silent spring that Carson predicted? It depends on the decisions that we make.

Think Questions

1. What environmental concerns have you have heard about in the news? Record a topic in your notebook and add any details you know.

2. What questions do you still have about this topic?

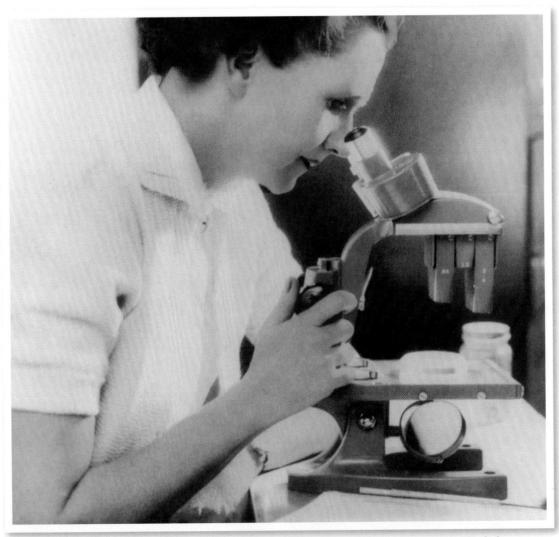

Regarded as the finest nature writer of the 20th century, it was Carson's training as a marine biologist that informed her writing.

A moose is a primary consumer. It eats only producers, such as the leaves and twigs of tall shrubs, and aquatic plants like water lilies.

Trophic Levels

A visit to a taiga forest is dominated by a lush population of producers, trees, shrubs, grasses, and mosses.

A small herd of moose can be seen munching on the greenery. Lurking in the shadows, a lone bear is looking for its next meal.

Different organisms play different roles in an ecosystem. These roles are called **trophic levels**. Trophic levels are feeding levels. The trophic level at the base of the ecosystem is the producers.

Trophic levels can be compared by their biomass. Biomass is the **mass** of matter produced by organisms in the ecosystem. Biomass includes all the dead and living organisms. The producer trophic level always has the largest biomass. The next-largest amount of biomass is the **primary consumers**. These consumers eat producers. The primary consumers are eaten by **secondary consumers**. This trophic level has little biomass, because there are fewer secondary consumers than primary consumers. The secondary consumers are eaten by **tertiary consumers**, which are even fewer. Each tropic level has less biomass than the level below it.

The purple lupines, grasses, and trees shown here are the base of the diagrammed food web. Directly or indirectly, all of the other organisms depend on them for energy.

Food Webs

A typical ecosystem has producers, consumers, and **decomposers**. Decomposers consume dead organisms and other biomass. All types of organisms transfer food energy in a food web. A food web is a model of the flow of energy and matter in an ecosystem. It shows all the feeding relationships. It also points out that some organisms must die to support other organisms.

The arrows in a food web connect organisms that eat other organisms. The arrow points from an organism to the organism that eats it. It points in the direction that the energy in the food goes. When a fox eats a chipmunk, the food energy in the chipmunk goes into the fox. The arrow points from the chipmunk to the fox.

A food web is a more accurate representation of the feeding relationships in an ecosystem. It rarely occurs that there is a one-to-one relationship between a **predator** and prey. Most consumers eat a variety of food organisms. Many consumers may compete for the same food source.

A Food Web

A food web shows the overlapping feeding relationships in an ecosystem.

Arranging organisms by trophic level helps show community interactions. In this food-web diagram, lupines, grass, and pine trees are producers. Gophers, voles, and chipmunks are primary consumers. They are **herbivores**, because they eat only plants. Barn owls, coyotes, and foxes are secondary consumers. Secondary and tertiary consumers are **carnivores**, or meat eaters. They eat other animals. Some animals, including most humans, eat both producers and consumers. We are **omnivores**.

Take Note

Look in your notebook for a food web that you made in class. How could you draw the food web to show the relative amounts of biomass of producers and consumers?

Energy Transfer

What happens to the food made or eaten by an organism? Much of the available energy maintains life functions. Some food energy is stored as biomass in the body of the organism. Most is used to do work, or passes back into the environment as thermal energy. So the energy needed to run, think, digest, pump blood, and perform all other life activities passes through the organism and into the environment.

Most energy enters the ecosystem as light from the Sun. It transfers to one or more organisms, and exits to the environment as thermal energy, commonly called body heat. This energy is transferred to the atmosphere, hydrosphere, or geosphere. It is not stored in biomass, and it is not eaten by the next trophic level. Only a small part of the food energy consumed at a trophic level is retained as biomass. Sometimes it is as low as 10 percent.

Scorpions are secondary consumers—carnivores—preying on insects, spiders, and even other scorpions. They can go up to a year without eating anything!

Consumers have to eat a lot just to maintain their body mass and functions. The biomass of producers in an ecosystem is always much larger than the biomass of primary consumers. And the biomass of primary consumers is much larger than the biomass of secondary consumers.

Trophic Pyramids

At each trophic level, energy transfers to new biomass through growth and reproduction. This energy is usually much less than the energy available in what organisms eat. That means the prey usually outnumber predators. For the same reason, there are usually fewer herbivores than plants. So trophic-level diagrams are often pyramids.

We arrange the ecosystem roles from producers, to primary consumers, to secondary consumers, and so on. The result is a layered diagram of the ecosystem.

Each layer is smaller as you go up the trophic levels. Each level of the **trophic pyramid** has less biomass than the one below it.

Where do we place the decomposers in the trophic pyramid? They interact with all the other trophic levels, so they are often placed along the side.

It is extremely rare for an ecosystem to have fourth-level consumers. So much energy is used or lost to the environment before each transfer that it is nearly gone after the third level. Fourth-level (and higher) consumers usually live in aquatic or marine ecosystems. An orca, for example, is a fifth-level consumer. An orca might eat a sea lion, which ate a salmon, which ate an anchovy, which ate zooplankton, which ate a single-celled alga. And then the great white shark eats the orca . . . You get the idea.

Animals that feed at several trophic levels are generalists. They should be represented in each trophic level where they play a role.

Take Note

Raccoons and crayfish both eat plant material, which makes them primary consumers. But they also eat insects and other animals, which makes them secondary or tertiary consumers. Where would you place them in a trophic pyramid?

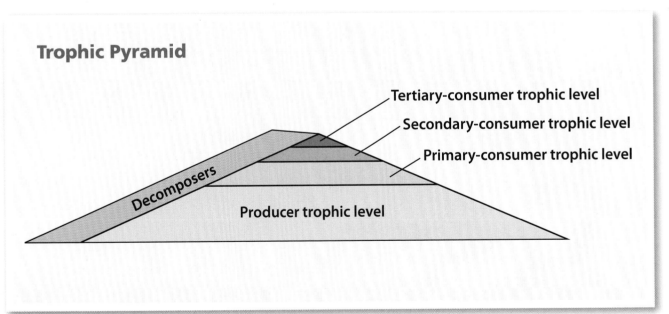

Trophic Pyramid

Tertiary-consumer trophic level

Secondary-consumer trophic level

Primary-consumer trophic level

Decomposers

Producer trophic level

The amount of energy and matter moving through a food chain is represented by a pyramid. It takes many organisms at the base to support just a few organisms at the top.

One Possible Food Chain

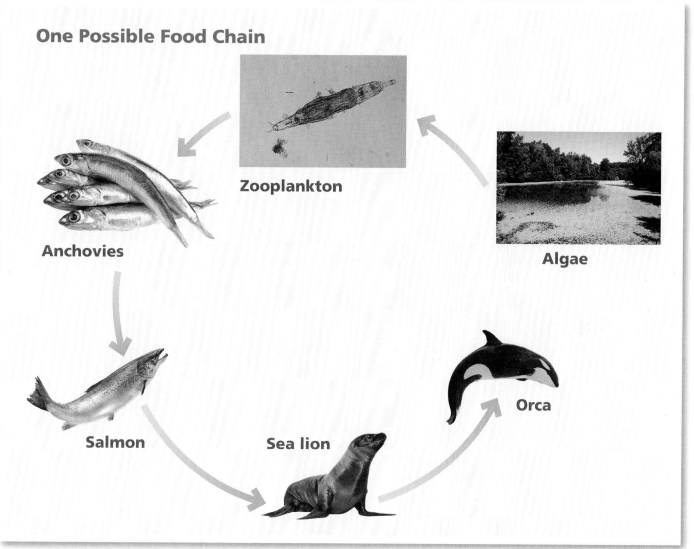

All food chains begin with producers, here algae, which use energy from the Sun to make their own food. A food chain with this many consumer levels is more common in aquatic ecosystems than terrestrial.

Pulling It All Together

Living organisms are complex. Their bodies are made of matter. The functions of living organisms are driven by energy. The energy for life comes from the Sun, captured in energy-rich molecules. The energy-rich molecules when consumed or used are called food.

Every organism needs a constant supply of matter and energy. Autotrophs (producers) get matter and energy from raw materials in the environment. Heterotrophs (consumers) get matter and energy by eating other organisms. Feeding relationships move matter and energy through the trophic levels in an ecosystem.

Globally, forests cover about one-third of Earth's land surface. As in all ecosystems, their producers make up the lowest trophic level and have the greatest biomass.

Dead organisms still have valuable matter and energy. Decomposers get the last bit of energy out of this organic material. They reduce the material to simple chemicals. Matter enters the ecosystem as food made by producers. It returns to the environment to be used again by living organisms. Matter recycles again and again. But energy passes through the ecosystem only once. Almost all the energy that passes through the ecosystem radiates into the environment as thermal energy. Once transferred to the environment, it is no longer useful as food energy.

Think Questions

1. **One model of an ecosystem is a trophic pyramid. Explain why the bottom layer of a trophic pyramid is the biggest, and the top layer is the smallest.**

2. **Why doesn't 100 percent of the energy pass from trophic level to trophic level?**

Decomposers

Many producers and consumers die without being eaten. What happens to their bodies? Deciduous trees lose their leaves every year. Where do the leaves go? Animals produce feces. What happens to them?

Dead organisms, dead parts of organisms, and feces are made of matter. This biomass collects on the ground, or the bottom of a lake or the ocean. It is available for other organisms to eat. Bacteria and fungi consume this organic matter. These consumers are known as decomposers.

Decomposers like mushrooms and other fungi get their energy by breaking down dead plant and animal matter.

When something rots or decays, it is actually being consumed by decomposers. This process of decay transfers every last bit of energy from the dead organism to the decomposers. The decomposers leave only a few simple chemicals, matter which becomes soil, water, and carbon dioxide in the air. The matter can be used once again by producers to make food during photosynthesis.

Bacteria and fungi are the most important decomposer organisms. They play a clear role in the ecosystem. But what about organisms like worms and termites, which eat dead plant matter for their energy? What about

Vultures eat the bodies of dead animals. Such animals are sometimes called scavengers. Do you think they should be considered decomposers or carnivores?

vultures and coyotes, which eat dead meat? Are these organisms decomposers, or are they herbivores and carnivores? Ecologists are still defining the roles of these important organisms that clean up the dead and discarded bits of life.

A film of dead organic matter covers much of Earth's surface. This organic matter, known as **detritus**, is the home of the **detritivores**, which eat dead material. Not all detritivores reduce the matter to simple chemicals. Worms and beetle larvae, for instance, leave behind a significant amount of matter. They produce feces, which are consumed by decomposers who can use the energy left in the feces.

Decomposers provide many ecosystem services. By recycling dead matter, they make important nutrients available to organisms. Without decomposers, the ecosystem's recycling crew, Earth would be buried under a layer of waste and dead organisms!

Think Questions

1. What does the following sentence mean? "Matter recycles again and again. But energy passes through the ecosystem only once."

2. What is the difference between a decomposer and a detritivore?

Maggots may seem disgusting, but they perform an essential role in helping make nutrients in an ecosystem available for organisms to use.

Limiting Factors

Nothing lives forever. Even the bristlecone pine trees in the western United States die after a few thousand years. Most organisms live much shorter lives.

Many insects live a few months. Fish and small mammals live a few years. Many plants, reptiles, birds, and large mammals live a few decades. A few others, like trees, may live a few centuries. Individuals always die, eventually. So for a species to survive, it must continually produce new individuals through reproduction.

A species has a **reproductive potential**. That is the rate at which its population can increase. Some species, like elephants, have low reproductive potential. A female elephant can produce a single offspring every 4 years. A female Atlantic cod, on the other hand, can lay 10 million eggs in 1 year. Clearly the cod has greater reproductive potential than the elephant.

So why aren't billions of trillions of cod filling up the Atlantic? Because there are **limiting factors** for every population on Earth. Abiotic and biotic limiting factors control the size of populations.

The bristlecone pine is one of the longest living species. In North America, the average age of these trees is around 1,500 years, with the oldest tree over 5,000 years!

Predator-prey relationships are biotic limiting factors that tend to keep populations in balance. Mountain lions affect a deer population through predation, and the availability of deer to hunt affects the size of the mountain lion population.

Biotic Limits: Predation

When one organism eats another, that is a limiting factor. Every organism is a source of food for some other organism. As you know, food provides the energy for life. When an individual reproduces, its offspring are new biomass. Predators take advantage of this new energy source. We see this kind of population control in Mono Lake. Brine shrimp feed on algae, reducing their numbers. In turn, the phalaropes and gulls eat brine shrimp, reducing their numbers. Predation can occur at any stage in the life cycle of an organism. When predators eat individuals, they limit the population size.

Humans are also predators. We collect fish by the boatload. We can harvest a field of wheat in a day. Sometimes we hunt just for sport. Free-ranging (feral) house cats are also efficient predators. The estimated 15 million feral cat population kills up to 4 billion birds and 22 billion mammals in the United States every year.

Biotic Limits: Disease

Diseases limit populations in the same way. We might not think of a bacterium preying on a large animal or plant. But the result can be the same. A mountain lion capturing a deer removes an individual from the population. A diseased organism entering a population can kill many individuals.

Humans can now quickly travel from a jungle to a city across the world. This speed helps spread infectious diseases. Diseases have become an important limiting factor for human populations. They also limit the many organisms that interact with us.

Did You Know?

Infectious diseases are caused by various microorganisms, including bacteria, viruses, fungi, and parasites. Microorganisms can be found in contaminated food or water. These organisms can also sometimes pass directly from one host to another during close contact.

Biotic Limits: Resources

Populations are limited by food supply. If an organism cannot get the energy to survive and reproduce, the population will decrease. Reproductive potential also decreases. If a snake cannot find and eat enough mice, it will starve to death. Even if it survives, it might be so weak that it cannot reproduce. Similarly, if there is a poor acorn crop, squirrels may starve. Even if they survive, they might not be able to feed their young. In 1982, the population of brine shrimp in Mono Lake was small. That year, the California gulls could not successfully feed their chicks. Most of the chicks died. Lack of food is one of the most important limitations on populations.

Biotic Limits: Competition

In an ecosystem, there can be several populations that eat the same foods. These populations will be in direct **competition** with each other. If one of the populations outcompetes other predators, it will reduce the amount of food available to others. This can limit other populations.

Bank swallows nest in colonies along the banks of streams.

Abiotic Limits: Reproductive Environments

Many organisms need particular conditions for reproduction. For example, the number and quality of nest sites limit bird reproduction. Bank swallows dig nesting burrows in sandy cliffs. If a cliff tumbles down during a flood or earthquake, their nesting sites are lost. Salmon lay their eggs only in clean gravel. Black bears give birth in winter dens. Without these physical conditions, no offspring will be born.

Reproduction is also limited by habitat availability. As humans clear forests and grasslands, we change the habitats where organisms live. Sometimes the habitats no longer allow certain populations to reproduce. Habitat destruction is a serious limiting factor for many organisms, especially those in danger of **extinction**.

Take Note

Review your observations of the classroom habitats. What are some limiting factors in each habitat?

Bank swallows made nesting burrows in this sandy cliff. Erosion or collapse may destroy such sites and limit their ability to reproduce.

In the past, bison would migrate hundreds of miles searching for food in winter. The bison habitat has been decreased by human land development, and the bison population is now tiny.

Abiotic Limits: Seasons

Seasonal changes affect populations. In temperate and polar latitudes, winter is a major limiting factor. Winter days are short, so production by photosynthesis slows or stops. Winter often brings rain, snow, and wind. Each of these can stress populations.

Some animals travel to avoid winter weather. Birds are famous for migrating to warmer regions. Other animals, like caribou, travel to find suitable winter environments. Some organisms become dormant, basically shutting down until spring. Frogs, fish, bears, squirrels, snakes, and maple trees use dormancy, reduced activity, and winter sleep. These strategies work under the right conditions: The wintering place has to give enough protection. The organism has to accumulate enough fat or food to survive the winter.

Winter is the main limiting factor for many temperate and polar populations. Many populations decrease to low levels, like the brine shrimp in Mono Lake. They expand rapidly in the spring. Such seasonal changes in population size are normal and healthy.

Abiotic Limits: Climate Change

Climate change is an important abiotic limiting factor for many species. Earth's climate always changes over time. But it has been changing more quickly over the last 100 years. Increased amounts of greenhouse gases in the atmosphere have caused this acceleration. These gases (including carbon dioxide) effectively trap thermal energy in Earth's atmosphere. The average global temperature is getting higher.

Climate change is predicted to speed up species extinction over the next 100 years. Mountain and arctic ecosystems are particularly sensitive to climate change. There will be less suitable habitat for animals that live there. One consequence of climate change is a rise in sea level. As it rises, salt water can seep into freshwater systems. Organisms will have to relocate, or might die.

Climate change can alter where species live and how they interact. These changes could transform ecosystems. As you know, changes in one species can ripple through the food web and affect many organisms. The climate influences timing for migration, blooming, and mating. As the climate changes, the timing of these events also changes. This can lead to missed connections with other organisms, limiting their populations. Growth and survival decrease when food sources are not ready when animals need them.

Sea ice loss due to climate change is threatening polar bear populations. Alarming melt rates are reducing their frozen habitats and access to prey.

Carrying Capacity

Every living thing has fundamental requirements for life. If it does not get those things, it dies. One of the most critical requirements is energy.

Energy enters the ecosystem as sunlight, or solar energy. Photosynthetic organisms capture some of the energy by transforming it into carbohydrates, and storing it. Limiting factors include light, space for living, and availability of water, carbon dioxide, and minerals. For any given ecosystem, producers can make a limited amount of food.

We know that the other organisms in an ecosystem acquire energy by eating each other. Primary consumers eat producers, secondary consumers eat primary consumers, and so on. The total number of consumers is limited by what producers can make.

Reaching Capacity

The total number of individuals that an ecosystem can sustain indefinitely is the **carrying capacity**. For instance, a backyard ecosystem has a certain amount of vegetation. The carrying capacity for rabbits could be three, year after year. If six rabbits move in, they might exceed the carrying capacity of the ecosystem. As a consequence, the rabbits could eat so much vegetation that they wipe out the producer populations.

Exceeding the carrying capacity of an ecosystem changes the nature of the ecosystem. Say the rabbits ate so many producers that the producers could not reproduce. The rabbits' food source would disappear. Baby rabbits would have nothing to eat, so the consumer population would die off. **Sustainable** ecosystems must have enough survivors of every species to reproduce and keep the population going.

Primary Production

Usually, the overall carrying capacity of an ecosystem depends on the amount of **primary production**. Primary production is the amount of food produced and redistributed throughout the food web during a year. The primary production of an ecosystem is a major part of its carrying capacity. Another limiting factor is how long it takes producers to make food. Consider the carrying capacity of Mono Lake. The lake has plenty of light, water, carbon dioxide, and minerals. The producers of the lake, the algae, reproduce rapidly. Food production in the lake is extremely high. The algae in the lake support trillions of brine shrimp and brine flies. These in turn nourish millions of birds and a few coyotes. A lot of life flows through Mono Lake each year.

The carrying capacity of rabbits in a backyard ecosystem is the stable population size that can be supported by the available resources. Too many baby rabbits would upset the balance.

Contrast this with the time needed to produce food on the Great Plains. Grasses grow more slowly than algae. They take longer to produce food. Grazing by insects, rodents, deer, bison, and cattle must not exceed the capacity of the grasses to produce more food.

The Great Plains and Mono Lake are both productive ecosystems. Each ecosystem has its own carrying capacity. It all depends on the ability of the producers to support the food web.

Think Questions

1. **What are some biotic limiting factors?**
2. **What are some abiotic limiting factors?**
3. **What are some limiting factors for your ecoscenario? Consider temperature, food sources, and other factors mentioned in this article.**

Even under ideal conditions, the Great Plains grasslands ecosystem takes a long time to produce the food that sustains its numerous consumer populations.

In winter, the surface of Mono Lake is still and the surroundings stark. But conditions are just right for an algae population explosion.

Mono Lake throughout the Year

Mono Lake appears dramatically different throughout the year. During some times, the surface of the lake appears desolate and devoid of life.

Come back at other times, and the lake is teeming with life above and below the water. Mono Lake is an ecosystem of contrasts.

Winter. During the cold of winter, Mono Lake looks like a wasteland. No birds swim on its surface, and no brine flies or brine shrimp are active. However, the lake is turning a healthy dull green color, as the planktonic algae experience explosive population growth.

The algae population increases rapidly because for a while there are no limits on its growth. Light, nutrients, and carbon dioxide are plentiful. The algae are not limited by the near-freezing temperature of the water, and most of the predators are dormant.

Spring. As spring arrives, the increased solar energy warms the water. Brine shrimp and brine flies hatch and start eating the algae. The hungry predators eat the algae faster than the algae can reproduce, so the algae population starts to decline.

Summer. As summer approaches, the first birds, the gulls, arrive. They feast on the brine shrimp. The California gulls slow the population growth of the brine shrimp, but do not reverse it. It is not until July, when all the migrating birds are present, that the brine shrimp begin to decline. They decline because they are reaching the limit of their food supply on one side, and hundreds of thousands of birds are eating them on the other side. The brine shrimp population declines rapidly.

In summer, swarms of brine flies are so thick at Mono Lake that they nearly hide the California gulls that are feeding on them.

Fall. By October, the nutrients in the water have been depleted, and predation has taken its toll on the algae population. It falls to its lowest level. The cooling water causes the eggs laid by the remaining brine shrimp and the pupae of the brine flies to fall into dormancy. The adult brine shrimp and brine fly populations drop to zero.

When the surface water approaches freezing, it is so dense that it plunges to the bottom of the lake. It stirs up the nutrient-rich sediments, making them available to the algae. The algae begin, once again, to reproduce rapidly, turning the lake green. The producers have set the stage for a repeat of the age-old surge of life in the Mono Lake ecosystem.

Think Questions

1. **Why do algae populations increase during winter months?**
2. **What causes the rapid decline of brine shrimp in the summer?**

Mono Lake changes color in spring as the algae are consumed by newly hatched and hungry predators: brine shrimp and brine flies.

Biodiversity

In 2016, more than 250 communities around the world joined together. They conducted a BioBlitz to count all of the species in their location.

In the United States alone, almost 7,000 participants identified more than 13,000 different species in just the national parks. Why count organisms?

Biodiversity is the variety of life. It refers to all the different kinds of organisms living in an area. It can be studied at many scales. Biodiversity can include the entire Earth and all living organisms, or just one part of the schoolyard.

Let's look at the biodiversity of a pond. At first glance, we see turtles sunning themselves, dragonflies zipping by, fish swimming. Getting closer, we notice small animals like leeches and water striders. Algae, cattails, and water lilies live in the pond. The tall grasses, shrubs, and trees next to the pond add to the biodiversity of this ecosystem.

Biodiversity is important because all the organisms in an ecosystem depend on other organisms within the ecosystem. At the pond, turtles eat dragonflies and vegetation, and they are food for the pond's herons and raccoons.

You need a microscope to see all of the pond's biodiversity. Thousands of different microorganisms live in the pond water.

Distribution of Biodiversity

Sampling is one way to measure biodiversity. We count the number of species living in an area. The sample is used to calculate a **biodiversity index**. The index is a measure of an ecosystem's health.

Typically, regions near the equator have the greatest biodiversity. This is true for both terrestrial and aquatic biomes. Tropical biomes are warm and sunny. They have abundant rainfall year-round. This climate provides energy and water in a stable environment for a huge number of species. Biomes that have severe weather (hot or cold) or little precipitation have less biodiversity. Fewer species can live in these environments.

Importance of Biodiversity

All species in an ecosystem are **interdependent**. When biodiversity is low and a population dies off, it can result in greater competition for the remaining resources. The entire ecosystem can fail. An ecosystem with high biodiversity is more likely to recover when a population disappears from it.

Did You Know?

Scientists have identified around 1.5 million species on Earth. One estimate of the number of species on Earth is about 8.7 million. Most unidentified species are microorganisms.

A keystone species fills a specific niche within an ecosystem and cannot be replaced by another organism. Wolves in Yellowstone National Park are keystone species that keep other populations from growing too large and causing negative effects throughout the ecosystem.

Ecosystems with high biodiversity can recover faster from natural disasters than the same kind of ecosystem with lower biodiversity. Remember that humans rely on ecosystem services such as food and clean water. So a high-biodiversity ecosystem that can recover quickly from an extreme storm, fire, or earthquake can best provide resources that humans need.

Not every species contributes equally to biodiversity. **Keystone species** are essential for the health of an ecosystem. Many keystone species fill a specific **niche**, or role, within the ecosystem. The role of the keystone species cannot be replaced by another species. Losing a keystone species starts a chain of events that affects the biodiversity of the entire ecosystem.

When keystone species are top predators, they control populations in a food chain. The wolves in Yellowstone National Park and sea otters in Monterey Bay National Marine Sanctuary are keystone species. They keep their prey populations from growing too large.

Bees and insects that pollinate plants are also keystone species. These pollinators are important for the good health of ecosystems. Bees, birds, and other creatures pollinate 75 percent of the world's food crops.

Ecosystem engineers are a special kind of keystone species. They change the ecosystem they live in. Beavers are ecosystem engineers. They build dams in rivers and streams, creating meadows. Earthworms, as detritivores, are also ecosystem engineers. By breaking down organic matter, they change the character of the soil they live in. That affects what grows there.

Threats to Biodiversity

The major threat to biodiversity is extinction. A species becomes **extinct** when its members no longer successfully reproduce. All of the organisms connected to the species are also affected. Organisms that ate the species have to find another food source or die. Extinction of a keystone species is a terrible blow to an ecosystem.

Extinction is part of life. About 99 percent of species that have ever existed on Earth are now extinct. There have been five major extinctions in Earth's history. These events were caused by meteor impacts, volcanic eruptions, and natural climate change. Scientists believe that a sixth major extinction is currently happening. This extinction is caused by humans. Some scientists think this marks a new geologic age. They call it the **anthropocene**. *Anthropo* means human.

Humans have deliberately affected the size of other populations on Earth for at least 10,000 years. Many of these changes have become more rapid in the past few centuries.

Scientists estimate that one-third of all known living species are currently threatened with extinction. Here are some key ways humans affect biodiversity.

Habitat loss. Change or destruction of habitat reduces biodiversity. Organisms that cannot survive outside their original habitat go extinct. Biodiversity decreases. Here are the main ways that humans change habitats.

- Construction
- Agriculture
- Deforestation (cutting down forests)
- Dragging fishing nets in the ocean
- Changing the flow of streams and rivers
- Draining and paving over wetlands

The Hoover Dam generates enough electricity for 1.3 million people every year, but has severely decreased fish populations.

Overhunting and overfishing.
Humans are predators. We eat other animals for food. Sometimes too many animals in a population are killed. This is called overhunting or overfishing. Overhunting can wipe out a population of food animals from an ecosystem, reducing its biodiversity.

Invasive species. Humans often introduce organisms into an ecosystem. These organisms can be aggressive and can change the ecosystem. These **invasive species** compete for resources with native organisms. They can decrease or wipe out populations, leading to a decrease in biodiversity. Many invasive species don't have a native predator to limit the population.

Pollution. Introducing chemicals or other substances into an ecosystem can harm organisms. Pollution can decrease the biodiversity of an ecosystem by damaging or killing organisms.

Climate change. Climate change affects temperature and precipitation patterns. Changes in these abiotic factors can have a big effect on the survival of local plants and animals. Another effect of climate change is rising sea levels. The rising ocean reduces land space for plants and animals. It can add salt to freshwater biomes. If populations in these ecosystems cannot survive the changes, biodiversity decreases.

Deforestation, often so that land can be developed for human use, drastically changes an ecosystem. The resulting habitat loss is a major threat to wildlife.

Protecting Biodiversity

Learning that humans are driving species into extinction is a heavy burden. But here is some good news. We can use science and technology to change our impact. We can have a positive effect on biodiversity. Our use of ecosystem services determines the size of the impact we have on an ecosystem, known as our **ecological footprint**. It should be our goal to leave a smaller footprint. Here are some ways to try to do that.

Think about what you eat. You might be able to eat lower on the food chain. What does that mean? Eating meat can affect an ecosystem more than eating plants. It takes a lot of energy and other resources to raise animals for food. You do not have to become a vegetarian. Just be more aware of what you eat. You can be more of an omnivore than a carnivore.

Many human activities help, rather than harm, the environment. Replanting local species is just one way to restore or conserve an ecosystem.

Reduce, reuse, recycle. We consume more than just food. We also consume a lot of natural resources, such as metal and wood. Getting these materials affects ecosystems. So it is a good idea to reduce their use. Consider recycling materials like cans and paper. Buy products made from recycled materials. Reusable grocery bags, water bottles, and coffee mugs reduce the use of new materials.

Think globally, act locally. It is easy to get overwhelmed by big environmental problems, like deforestation and climate change. Remember that every world leader started by seeking local solutions. Consider small ways to improve your backyard, schoolyard, and neighborhood. Volunteer with a local organization, or start your own. Roots and Shoots is a great online resource to help you get started.

 For more information on the Roots and Shoots program, visit FOSSweb.

Think Questions

1. **Why is biodiversity important to an ecosystem?**
2. **What things can you do to reduce your ecological footprint?**
3. **What programs does your community have to reduce its ecological footprint?**

Recycling is a great first step in conserving our resources and reducing our ecological footprint.

Invasive Species

In 1876, an English banker visiting America grew fascinated by the North American gray squirrels. He took two squirrels home to England with him, and released them in his yard.

Little did he know that this squirrel population would wreak havoc across Europe over the next century. A **native species** is a kind of organism that has been part of an ecosystem for a long time. **Introduced species** are brought in by humans. They become part of the local ecosystem and have several names. We refer to them as introduced species, but they can also be called nonnative or exotic species.

Introduced species can change the balance of an ecosystem. Not all introduced organisms survive in their new home. And not all introduced species have negative effects on native species. But when they do, it can be disastrous. When an introduced species has a negative impact on an ecosystem, it is labeled an invasive species.

The European starling is an invasive species that has grown from 60 introduced birds to an estimated 200 million. Flying in flocks of thousands, they devour crops, compete with native songbirds, leave disease-carrying droppings, and pose a hazard to air travel.

How Invasive Species Travel

How are organisms introduced to new ecosystems? Humans sometimes introduce them on purpose. In 1890, 60 European starlings were released in New York City during a Shakespeare festival. Without natural predators, the birds survived and flourished. Today, millions of starlings live in North America. They cause problems for native birds by competing for nest sites and resources.

Most invasive species are accidentally introduced. Many aquatic invasive species hitched a ride on the bottom of a ship or in ballast tanks (water tanks used to stabilize a ship). Trading of natural goods transports many species. Timber, plants, and pets provide hiding places in their bark, leaves, and fur.

Take Note

What invasive species have you heard of? List them in your notebook.

Impacts of Invasive Species

An invasive species can change habitats. It can alter ecosystem functions and services. Invasive species are a bigger problem than overhunting, pollution, and disease combined. In the United States, interactions with invasive species are the biggest threat for half of the species at risk of extinction. Invasive species threaten biodiversity and ecosystem services in these ways.

- Crowding out native species
- Interfering with human activities, such as raising livestock or crops
- Transmitting disease
- Acting as predators
- Changing existing habitats

Major US Invasive Species

Powerful predators. Introduced predators can throw off the balance of the food chain. If they are too successful, they can reduce or wipe out other populations. These problems cause other changes throughout the ecosystem.

The *black rat* is native to tropical Asia. Rats hitched rides on ships as long ago as 100 CE. They thrived in the growing cities where ships would dock. These predators will eat nearly anything. They seem to enjoy bird and reptile eggs and baby animals. They might have caused the extinction of many species of birds, reptiles, and other small animals. The rats also carry fleas that can infect humans

"Black Death" plague epidemics decimated Europe's population in the 1300s. Black rats with plague-infected fleas carried the disease westward along trade routes from their native Central Asia.

Scientists estimate that nearly 50 million pet cats may be outside hunting, along with an equal number of feral (stray) cats.

with disease. The plague, transmitted by fleas on rats, killed 75 to 200 million people in the 1300s.

Free-ranging *domestic cats* are a global threat to small-mammal and bird biodiversity. In the United States alone, domestic cats kill up to 4 billion birds and 22 billion mammals every year. Recent studies suggest that domestic cats kill more native birds and mammals in the United States than any other human impact.

Did You Know?

Archaeological evidence suggests that cats were domesticated by humans as early as 10,000 years ago in northern Africa. Their hunting abilities are often considered a valuable asset to keep homes rodent-free.

Guam now suffers from an infestation of brown tree snakes. The invasive predator has silenced the island's forests by wiping out most of the birds.

The *northern snakehead fish* seems like the perfect invasive predator. Native to the waters of Southeast Asia, this fish will eat anything it finds. It took only a few individuals released into the eastern US watersheds to cause trouble. What makes them so dangerous to biodiversity is that they can survive outside of water for several days. Each mature female can lay up to 75,000 eggs a year. Young fish can travel on land to search for a new home!

The *brown tree snake* accidentally traveled to the island of Guam in cargo shipments. Within 20 years, the introduced snake population eliminated 10 out of 11 bird species native to the forests of Guam.

Did You Know?

Brown tree snakes are rugged! These snakes can survive for an extended length of time without food. It has been reported that a brown tree snake survived for almost a year packed up in a used washing machine. The snake was 2.5 meters (m) long.

Lionfish came from the western Pacific and Indian Ocean. They were popular aquarium fish. As they grew too large, some owners released them off the Florida coast. With no natural predators and plenty of fish to eat, the lionfish population quickly grew. Some areas now have contests to encourage lionfish fishing. Marine sanctuaries want snorkelers to report seeing lionfish, to help prevent population growth. Some areas of the Caribbean are developing lionfish recipes to encourage people to fish for them.

Competitors and harassers. Some introduced organisms outcompete native organisms for food or habitat space. *Zebra mussels* traveled to the United States from Russia in ballast. With no natural predators, zebra mussel populations have grown dramatically. The mussels cluster on underwater surfaces. They clog pipe systems, causing billions of dollars in damages each year. Competition with the zebra mussel has driven at least 30 species of native freshwater mussels close to extinction. The Great Lakes' ecosystems have been significantly changed by zebra mussels.

Kudzu is a climbing vine native to Japan and China. It grows out of control in the southeastern United States. This fast-growing plant was introduced to prevent soil erosion. Without any natural predators, kudzu now covers and threatens many ecosystems.

Zebra mussels have spread from their Great Lakes "invasion" site to 29 states on boats moving through the Mississippi River basin.

The North American *gray squirrel* is a native species in deciduous forests in the eastern United States. It was introduced to England in 1876, and it now lives in deciduous forests across Europe. This squirrel is larger and more aggressive than European red squirrels. It outcompetes them in getting food. It is driving the native squirrels to extinction.

Microscopic invaders. Microscopic organisms can also be dangerous in a new ecosystem. *Chestnut blight fungus* came to the United States on trees imported from Asia. In the early 1900s, the fungus attacked American chestnut trees. It killed nearly 4 billion trees in 50 years. It is estimated that fewer than 100 mature trees survive in their eastern US ecosystem. The near-extinction of the American chestnut was a disaster for the many animals that depended on the tree for habitat or food, including white-tailed deer, wild turkeys, and black bears.

The Indian mongoose was introduced to New Zealand and elsewhere to control rodents and snakes. Unfortunately, it has preyed on native reptiles, birds, and mammals in many of its new homes.

114

Minimizing Impacts on Biodiversity

The total economic impact of invasive species in the United States is hard to calculate. They result in billions of dollars of lost crops. Efforts to combat them also cost billions of dollars a year. We use three main ways to minimize the effects of invasive species.

Prevention. Keeping an invasive species out is the best strategy to protect an ecosystem. Humans are the main transporters of invasive species. So education and regulations can be helpful. Many boat licenses require owners to clean their boat before launching it in a new body of water.

Border inspectors look for invasive species and materials that could carry hitchhikers.

Eradication. Say an invasive species has arrived in an ecosystem. The best strategy is to wipe it out before it becomes uncontrollable. Early efforts can reduce the effect on the ecosystem.

Sometimes humans introduce other organisms that eat or compete with the invasive population. However, introducing another organism can cause even more troubles. For example, the Indian mongoose was introduced to control rat and opossum populations in New Zealand. The mongoose hunted the native birds instead. Today, New Zealand has three invasive predators instead of two.

Mitigation. Once an invasive species has settled into an ecosystem, the only strategy is trying to minimize its effects. A lot of money goes to protecting native species, crops, and livestock.

What Can You Do?

You can help support native species in your area.

Plant a native garden. Many invasive plants were brought to the United States for gardens. People like to make their gardens look like other places around the world.

We can plant instead the many beautiful native plants that provide food and shelter for native animals. There are other benefits, too. Because native plants are well adapted to their ecosystems, they do not need excessive watering or pesticides.

Practice prevention strategies. Avoid taking plants, fruits, or animals to new places, especially when signs warn against it.

Keep your cat indoors. For most Americans, the best way to protect native biodiversity is to keep the family cat indoors. Studies show that putting a bell on a cat's

Dahlias are native to Mexico, where they have grown for centuries. They are now a popular garden flower worldwide.

collar is not helpful. Most wildlife will not identify the sounds with danger until it is too late. You can also help by supporting your local animal shelter. Many shelters help find homes for feral cats, so they hunt less. Shelters also curb cat populations by spaying female cats.

Support habitat restoration. You can help restore natural areas. Restoration projects can include removing invasive species, planting native species, and removing pollutants. When we minimize human impact on an area, we give native plants and animals a better chance of surviving.

Think Questions

1. Why is it a good idea to plant only native species in your garden?
2. Is it a good idea to release your classroom organisms into the local ecosystem? Explain your answer.

Mono Lake in the Spotlight

Life depends on water in Mono Lake, as it does in every ecosystem. Humans use a tremendous amount of water. Whenever we can use it for human use, we usually do.

Water Use at Mono Lake

The Los Angeles Department of Water and Power (LADWP) looked at the streams running out of the Sierra Nevada into what seemed to be a useless, salty lake. They decided to build dams to divert the water to Los Angeles before it went into Mono Lake. The project was completed in 1941. Almost all of the water that had flowed into Mono Lake was piped to Southern California. Without the annual inflow of fresh water, the lake began to dry up. The surface of the lake was at 1,956.5 meters (m) elevation in 1941. By the 1960s, the lake surface had dropped by 6.5 m.

By the mid-1970s, the lake level was down more than 12 m. The lake held half the amount of water it had in 1941. Changes started to appear in the ecosystem. The salt was twice as concentrated. This salinity change limited the populations of primary consumers, the flies and brine shrimp. The shrimp were not growing as large as usual, and their numbers were declining.

When Mono Lake's inflowing water was redirected to meet the growing water demands of Los Angeles, the ecosystem began to collapse.

118

In 1982, brine shrimp production was so low that the 50,000 breeding California gulls could not catch enough shrimp to feed their chicks, and 25,000 half-grown chicks starved to death. Furthermore, the water was so low that a land bridge connected Negit Island to the mainland. The bridge allowed coyotes and other predators to walk to the nesting area. They ate the gull eggs and chicks and drove the adults away.

A Plan for Restoration

In 1978, David Gaines (1947–1988), an expert in ecology and environmental advocacy, became concerned about the rapidly changing conditions in the Mono Lake ecosystem. He founded an action group called the Mono Lake Committee. He started to work on ways to reverse the damage to the Mono Lake ecosystem. He worked tirelessly with the National Audubon Society, California Trout, California Department of Fish and Game, US Forest Service, and LADWP to find solutions to the problem.

In 1994, after years of negotiations, the California Supreme Court ruled that the LADWP should release water to restore the lake. Over about 20 years, the new water level would reach 1,948 m and stay there. This was not as high as the lake used to be. It was a compromise between the needs of the ecosystem and the needs of people. With this compromise, Los Angeles gets some water, and the Mono Lake ecosystem is sustainable. The land bridge to Negit Island disappeared, trout can spawn in the creeks feeding Mono Lake, and visitors enjoy excellent views of Mono Lake.

David Gaines started and led the grassroots efforts that have saved Mono Lake.

The lake today is very close to the goal of 1,948 m. The added fresh water has returned salt concentration to historical levels. Water use remains a planning issue for Mono Lake managers and the LADWP. In Los Angeles, much of the water flowing to Mono Lake is not missed because the city has improved water conservation and recycling and conservation education. In one successful move, the LADWP replaced old toilets throughout the city with new low-flush models. They saved more water in 1 year than was ever diverted from Mono Lake!

Conserving Water in Your Ecosystem

Some places in the United States have more limited water resources than others. Water might not be a limited resource in your area. But simple water-conservation habits are a responsible way to help support your local ecosystem.

Showers. Showering is one of the main ways we use water at home. The average showerhead uses 10 liters (L) of water per minute. Installing a water-saving showerhead does not change the feel of the shower. And it can save your family 11,600 L of water a year! You can save another 800 to 1,200 L of water a year if you collect the cold water as you wait for your shower to warm up and use it to water your garden.

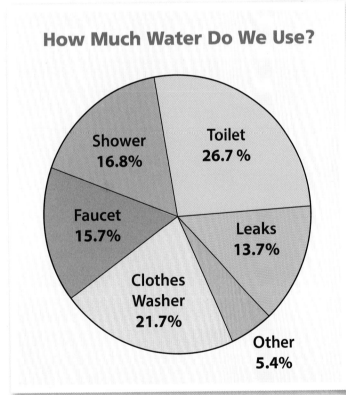

How Much Water Do We Use?

- Toilet 26.7 %
- Shower 16.8%
- Faucet 15.7%
- Clothes Washer 21.7%
- Leaks 13.7%
- Other 5.4%

This chart might help you think about where your family could use less water at home.

Faucet. The average faucet flows at 8 L per minute. If you could prevent 1 minute of water flow a day, that would be over 2,800 L a year! This strategy can be as simple as turning off the water while brushing your teeth or scrubbing dishes. The more thoughtful you are about turning the water off and on, the more water you can save.

Dishwasher. The average dishwasher uses just 16 to 24 L of water for a full washing cycle. Scraping food off dishes with a spatula instead of rinsing helps save water.

Drinking water. Many people run the tap until it is cool for drinking. Instead, fill a bottle with tap water and store it in the refrigerator to keep it cool. That way, you can save 800 to 1,200 L of water a year.

Running water only when it is being used may seem insignificant, but it's not. Turning off the faucet is a huge water-saver!

Toilet. Encourage your family to install low-flush toilets, which use only 6 L per flush.

Gardening. Use plants that are native to your area in the garden. They often need little maintenance and water, and support local wildlife. If you have a lawn, water it in the morning or evening instead of the hottest time of the day. This timing saves water that would be lost to evaporation. Another way to lower evaporation is to avoid watering on windy days. A layer of mulch around trees and plants can also reduce evaporation. You can also use a rain barrel to collect water that runs off your roof. It stores rainwater until your garden needs it.

Think Questions

1. **List ways you used water today.**
2. **How can you and your family save water?**

These barrels collect rain that can be used later for the lawn and garden. Collecting rainwater lowers the demand on community water resources; it also prevents erosion and reduces runoff pollution.

Images and Data

Images and Data   Table of Contents

Minihabitat Organisms

Earthworm

Genus: *Lumbricus*

Species: Varies

Size: Up to 25 cm long

Range: Worldwide except polar regions

Natural History: Earthworms live in the upper layers of the soil. They can tunnel as deep as 2 m if conditions are dry or cold. They prefer light, loamy soils to soils with a lot of clay and sand. Temperatures of about 13°C are ideal.

Food: Earthworms eat detritus. They decompose the organic material and return nutrients to the soil.

Predators: Birds, frogs, toads, salamanders, lizards, shrews, minks, raccoons, and turtles

Shelter: Tunnels in the upper layers of soil

Reproduction: Earthworms are hermaphroditic (have both male and female sex organs). The mating pair fertilizes each other. The resulting cocoon contains eggs. It is left behind in the soil when the pair separates. Tiny earthworms emerge in 2–4 weeks.

Abiotic Interaction: The movement of earthworms helps break up, loosen, and mix topsoil. Earthworm waste products enrich the soil and recycle nutrients into the environment.

Biotic Interaction: Humans add earthworms to gardens to improve the quality of the soil. Earthworms break down detritus and improve the soil. Earthworms are also used as fish bait.

Aquatic snail

Genus: *Planorbis*

Species: Varies

Size: Shell up to 3 cm diameter

Range: Temperate and tropical freshwater ponds

Natural History: Snails have a hard, spiraled shell. It gets bigger toward the opening as the snail grows. The muscular part that sticks out from the shell is the foot. Aquatic snails scrape algae from the surfaces that they travel on.

Food: Algae, aquatic plants, and detritus

Predators: Large fish, birds, waterfowl, shrews, and turtles

Shelter: If threatened, an aquatic snail pulls into its shell. It closes the opening with a door made of hard material, called an operculum.

Reproduction: Snails are hermaphroditic (have both male and female sex organs). During mating, two snails exchange sperm, and both can lay eggs. Eggs are laid on plants in a jelly capsule. The eggs hatch into tiny larvae that swim freely until they begin to grow a shell. The shell weighs them down, and they begin a life of crawling along the pond bottom.

Abiotic Interaction: When the water lacks calcium or is too acidic, snail shells can become weak. Changes in water quality from human activities (such as runoff of fertilizer into ponds or streams) can directly affect the snails' habitats.

Biotic Interaction: Aquatic snails are kept in home aquariums to reduce algae and detritus. They eat whatever food is left over by fish.

Isopod

Genus: *Oniscus*

Species: Varies

Size: About 1.8 cm

Range: Temperate and tropical regions

Natural History: Isopods have an exoskeleton. They live in dark, damp areas, especially under rocks and in leaf litter.

Food: Detritus, fruit, fungi, and young plants

Predators: Birds, frogs, lizards, turtles, and salamanders

Shelter: Under rocks and logs, and in leaf litter. Some species roll into a tight ball if threatened.

Reproduction: Females carry eggs until they hatch.

Abiotic Interaction: Isopods must have a dark, moist habitat. They enrich soil by recycling nutrients.

Biotic Interaction: Isopods break down detritus and can be a minor agricultural pest by eating plants.

Guppy

Genus: *Poecilia*

Species: *reticulata*

Size: About 3 cm

Range: Freshwater streams of Central America and northern South America

Natural History: Guppies are small fish that bear live young. Females are usually beige or silver gray. Males are smaller and have longer, flowing tails.

Food: Detritus, algae, small fish, and plants

Predators: Adult guppies eat unprotected young. In their natural environment, larger fish of other species prey on guppies.

Shelter: Females and young hide in vegetation. Guppies can live in an unheated aquarium unless the room temperature goes below 15°C.

Reproduction: Females can store sperm and can produce several broods from one mating. Babies are born live and must immediately find shelter to avoid being eaten.

Abiotic Interaction: Guppies are freshwater fish, so they cannot survive in a saltwater habitat.

Biotic Interaction: Brightly colored guppies are bred for home aquariums. Feeder guppies are raised as food for larger aquarium fish. Guppies released from aquariums can disrupt the predator/prey interactions in aquatic environments.

Scud

Genus: *Gammarus*

Species: Varies

Size: 5–30 mm

Range: Fresh water in the Northern Hemisphere

Natural History: Scuds are crustaceans related to isopods and shrimp. They are much more active at night than during the day. They crawl and walk using their legs, in addition to flexing their body.

Food: Bacteria, algae, and detritus

Predators: Fish, toads, salamanders, waterfowl, and other crustaceans

Shelter: Detritus or plant material close to the bottom or among submerged objects

Reproduction: Most scuds breed between February and October. Eggs and young develop in a brood pouch on the female. Young stay in the pouch about a week or so.

Abiotic Interaction: Scuds are freshwater organisms, so they cannot survive in a saltwater habitat. Scuds are found in almost all freshwater aquatic environments.

Biotic Interaction: Scuds help recycle detritus into the environment.

Tubifex worm

Genus: *Tubifex*

Species: Varies

Size: Up to 4 cm long

Range: Temperate freshwater ponds and streams

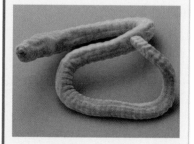

Natural History: Tubifex worms live on the bottom of ponds with their heads stuck into the substrate and tails waving in the water.

Food: Bacteria and detritus

Predators: Fish, amphibians, and crustaceans

Shelter: Soil and gravel at the bottom of ponds

Reproduction: Tubifex worms are hermaphroditic (have both male and female sex organs).

Abiotic Interaction: Tubifex worms are found in the mud of many natural waterways. They can survive in polluted water that other organisms find toxic.

Biotic Interaction: Tubifex worms are raised for live food for tropical aquarium fish. In waterways, they are detritivores.

Land snail

Genus: *Helix*

Species: Varies

Size: Shell up to 3 cm diameter

Range: Worldwide except polar regions

Natural History: Land snails have two sets of tentacles on their heads. The upper set contains nerve cells that are sensitive to light and smell. The two smaller tentacles are sensitive to touch. These lower tentacles can detect food, other snails, and surfaces. Snails get body heat from their surrounding environment. They become inactive if too cool.

Food: Detritus, plants, and calcium sources

Predators: Birds, skunks, and raccoons

Shelter: Snails live on the ground. They prefer cool, damp places. To escape dry conditions, they retreat inside their shells. They form a crusty layer over the opening to preserve the moisture in their bodies. This semi-hibernation is called estivation.

Reproduction: Snails are hermaphroditic (have both male and female sex organs). During mating, two snails exchange sperm; both can lay eggs. Eggs are laid underground.

Abiotic Interaction: Absence of calcium in the environment causes weak shells. Land snails can be found under detritus, in cool, damp places.

Biotic Interaction: Snails are detritivores, but are often considered garden and agricultural pests. For that reason, the US Department of Agriculture controls the movement of *Helix* across state lines. Some species of snails are eaten by humans as a delicacy.

Elodea

Genus: *Elodea*

Species: *canadensis*

Size: Up to 1 m long

Range: Throughout North America

Natural History: Elodea grows in freshwater ponds and slow-moving streams. It is a member of the tape-grass family.

Food: Photosynthesis

Predators: Fish, insects, aquatic snails, crayfish, turtles, and salamanders

Shelter: Roots in mud in quiet water

Reproduction: Elodea often reproduces vegetatively (a form of asexual reproduction). It drops winter buds, which root in the spring. It also produces small flowers at the tip, with seeds in small green capsules.

Abiotic Interaction: Productivity depends on light, water, and temperature levels.

Biotic Interaction: Elodea provides a protective habitat for many microorganisms. Elodea is a popular plant for home aquariums. Populations that are introduced to the wild can reproduce quickly and clog natural waterways. The sale of elodea is banned in some states because it is an invasive plant.

Duckweed

Genus: *Lemna*

Species: *minor*

Size: About 0.5 cm diameter

Range: Temperate freshwater ponds and lakes worldwide

Natural History: Duckweed is tiny, but huge populations can cover the entire surface of ponds and lakes. It is the smallest known flowering plant and a member of the duckweed family.

Food: Photosynthesis

Predators: Fish, aquatic snails, birds, water rats, and turtles

Shelter: Still surface water

Reproduction: Duckweed usually reproduces vegetatively by budding. Flowers rarely appear and produce seeds.

Abiotic Interaction: Productivity depends on light, water, and temperature levels. A population of duckweed can quickly cover the surface of calm waters.

Biotic Interaction: Duckweed is food for many aquatic organisms. It is sold in aquarium stores for home aquariums. Duckweed can be eaten by many small fish in aquariums.

Alfalfa

Genus: *Medicago*

Species: *sativa*

Size: 30–90 cm tall

Range: Temperate grasslands

Natural History: Alfalfa has a high tolerance for drought, cold, and heat. It has a taproot that can grow as deep as 15 m. It is a member of the pea family.

Food: Photosynthesis

Predators: Insects, birds, rodents, deer, and grazing livestock

Shelter: Mature plants reach a height of 1 m and can cover fields, providing shelter for insects.

Reproduction: Small bunches of flowers at the ends of the stems develop into coiled seedpods.

Abiotic Interaction: Productivity depends on light, water, and temperature levels. Alfalfa improves soil quality, so it is grown in alternating years with other crops.

Biotic Interaction: Flowers are pollinated by insects, primarily bees. Alfalfa is cultivated as a pasture crop or for hay.

Rye grass

Genus: *Lolium*

Species: Varies

Size: 1–2 m tall

Range: Temperate grasslands of Europe, Asia, Africa, and North America

Natural History: Rye grass grows in tufts. This grass is native to Europe, Asia, and Africa, but is now grown worldwide. It is a member of the grass family. It should not be confused with rye grain used to make flour for rye bread.

Food: Photosynthesis

Predators: Birds, insects, rodents, deer, and grazing livestock

Shelter: Rye grass grows in open, sunny fields. Left uncut, some species can grow nearly 1 m tall.

Reproduction: Rye grass reproduces by seed and is pollinated by the wind.

Abiotic Interaction: Productivity depends on light, water, and temperature levels.

Biotic Interaction: Rye grass is an important food for grazing livestock and is used for ornamental lawns, including grass tennis courts.

Wheat

Genus: *Triticum*

Species: Varies

Size: 30–90 cm tall

Range: Grasslands worldwide

Natural History: Wheat grows best in a temperate climate with rainfall between 30 and 90 cm per year. It is a member of the grass family.

Food: Photosynthesis

Predators: Birds, insects, rodents, deer, humans, and grazing livestock

Shelter: Wheat is grown in the sunny grasslands of the Midwest United States. Wheat is usually harvested when the plants are 1–2 m tall.

Reproduction: Wheat reproduces by seed and is pollinated by the wind. Plants live for one growing season.

Abiotic Interaction: Productivity depends on light, water, and temperature levels.

Biotic Interaction: Wheat is an important food, grown throughout the world. More land is devoted to growing wheat than to any other grain.

Milkweed-Bug Hatching Investigation

A class of middle school students decided to find out what variables affect the hatching of milkweed-bug eggs. They planned controlled experiments to help them understand milkweed-bug egg hatching. They gathered the data and organized it for others to share. They did not have time to summarize their results or to write their conclusions. The report is on the following pages.

Title. What Variables Might Affect Milkweed-Bug Egg Hatching?

Purpose. All organisms have limits on their populations. One limit on a population of milkweed bugs might be egg hatching. We decided to test three variables to see how they affect (1) the number of eggs that hatch and (2) the length of time before eggs hatch. The three variables we tested are temperature, humidity (moisture in the air), and exposure to light.

Experimental design. We started with a large habitat of breeding milkweed bugs. One day before the experiments, we put fresh pieces of polyester wool in the habitat. The next day, we had several thousand new eggs to use in our experiment.

We used special equipment to control the variables for the experiments.

A. A *temperature-control device* let us maintain precise temperatures.

B. A *humidity-control device* let us maintain precise humidity.

C. A *light-control device* let us maintain precise exposure to light.

The standard hatching environment was 25°C, 50 percent humidity, and 12 hours of light exposure a day.

We put 100 milkweed-bug eggs in each experimental setting.

- In the temperature experiment, only temperature changed. Humidity (50 percent) and light exposure (12 hours a day) were controlled.

- In the humidity experiment, only humidity changed. Temperature (25°C) and light exposure (12 hours a day) were controlled.

- In the light experiment, only light exposure changed. Humidity (50 percent) and temperature (25°C) were controlled.

We observed the eggs every 5 days. We also recorded the number of eggs that had hatched. We moved the nymphs to another habitat. Then we returned the unhatched eggs to the experimental conditions. The experiments continued for 30 days.

Data

Effect of Temperature on Milkweed-Bug Egg Hatching

Elapsed time in days

Temperature (°C)	0 days	5 days	10 days	15 days	20 days	25 days	30 days
0°	0	0	0	0	0	0	0
5°	0	0	0	0	0	0	0
10°	0	0	0	10	23	26	28
20°	0	11	86	91	91	91	91
30°	0	36	94	95	95	95	95
40°	0	57	92	92	92	92	92
50°	0	0	0	0	0	0	0
60°	0	0	0	0	0	0	0

Effect of Humidity on Milkweed-Bug Egg Hatching

Elapsed time in days

Humidity (%)	0 days	5 days	10 days	15 days	20 days	25 days	30 days
0%	0	26	80	96	96	96	96
25%	0	22	88	91	91	91	91
50%	0	28	90	95	95	95	95
75%	0	26	86	95	95	95	95
100%	0	21	87	96	96	96	96

Effect of Light Exposure on Milkweed-Bug Egg Hatching

Elapsed time in days

Light (hr/day)	0 days	5 days	10 days	15 days	20 days	25 days	30 days
0	0	28	88	94	94	94	94
6	0	22	83	94	94	94	94
12	0	25	90	97	97	97	97
18	0	23	82	91	91	91	91
24	0	26	88	96	96	96	96

Results. Look at the results for each experiment: temperature, humidity, and light exposure. Which conditions could be considered limiting factors? Use data to support your claims.

Conclusions. What do the results suggest about the limiting factors for milkweed-bug populations in nature?

Algae and Brine Shrimp Experiments

Purpose. Lab experiments were set up to determine if the abiotic factors of light and temperature limit population growth of algae and brine shrimp.

Experimental design. Four populations of planktonic algae and four populations of brine shrimp were placed in controlled environments. Population sizes were measured once a month for a year.

Algae experiments. Four identical aquariums were set up. Each had the same amount of Mono Lake water, nutrients (including carbon dioxide), and a small starter population of algae.

Two aquariums were maintained at constant temperatures (one at low temperature and one at high temperature), and the light was varied (changed).

The other two aquariums were maintained with constant light (one at low light similar to winter and one at high light similar to summer), and temperature was varied.

Algae Experimental Setup

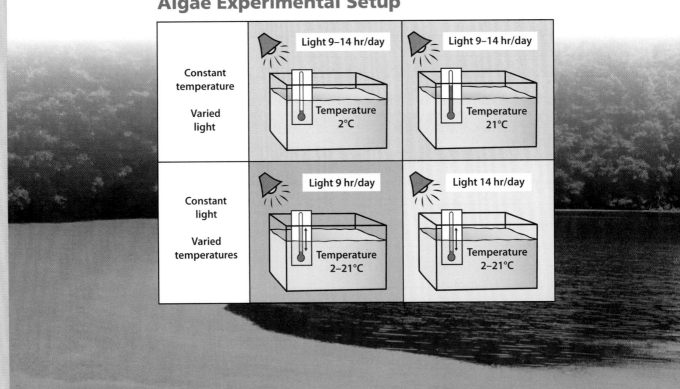

Brine shrimp experiments. Four identical aquariums were set up. Each had the same amount of Mono Lake water, food (powdered algae), and 1 g of brine shrimp eggs.

Two aquariums were maintained at constant temperatures (one at low temperature and one at high temperature), and the light was varied.

The other two aquariums were maintained with constant light (one at low light and one at high light), and temperature was varied.

Brine Shrimp Experimental Setup

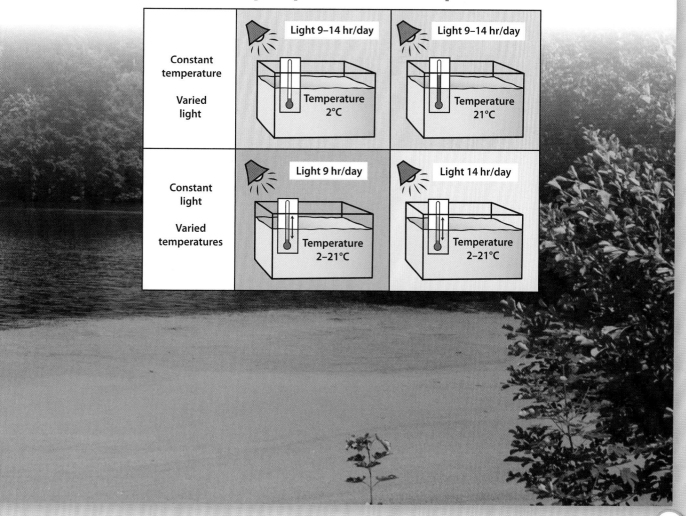

Experimental procedure

After they were set up, the eight aquariums were allowed to develop for 1 year.

Light

- The low-light aquariums received 9 hours of light a day. Nine hours of light represents the shortest days of the year at Mono Lake.

- The high-light aquariums received 14 hours of light a day. Fourteen hours of light represents the longest days of the year at Mono Lake.

- The variable-light aquariums received the amount of light that corresponds to the length of the calendar day at Mono Lake—9 hours in January, gradually increasing to 14 hours in June and July, then decreasing to 9 hours in December.

The light and temperature controls on the student aquariums model conditions at Mono Lake over the course of a year.

Temperature

- The low-temperature aquariums were 2°C all year. Two degrees is the lowest temperature of Mono Lake in the winter.

- The high-temperature aquariums were 21°C all year. Twenty-one degrees is the highest temperature of Mono Lake in the summer.

- The variable-temperature aquariums started out cold (2°C) in January, warmed gradually to 21°C in July and August, and then cooled to 2°C by December.

Populations were checked once at the end of every month. A sample of aquarium water was removed to measure chlorophyll a. The amount of chlorophyll a is reported in micrograms per milliliter (μg/mL). It indicates the size of the algae population.

Populations of brine shrimp were counted by placing a sample of aquarium water under a microscope and counting all the shrimp of any size (larvae, juveniles, and adults). The result was converted to the number of brine shrimp in thousands per cubic meter (m³) of water.

Results

Planktonic algae experiments (Algae population in µg/mL)

Temperature (°C)	Light (hr/day)	Jan	Feb	Mar	Apr	May	Jun	Jul	Aug	Sep	Oct	Nov	Dec
2	9–14	1	3	27	68	86	91	92	96	94	92	93	95
21	9–14	1	5	33	88	90	91	94	97	96	97	98	94
2–21	9	1	2	10	31	55	82	92	96	94	89	92	93
2–21	14	1	3	19	43	86	91	94	97	96	97	98	94

Brine shrimp experiments (Brine shrimp in thousands/m³)

Temperature (°C)	Light (hr/day)	Jan	Feb	Mar	Apr	May	Jun	Jul	Aug	Sep	Oct	Nov	Dec
2	9–14	0	0	0	0	0	0	0	0	0	0	0	0
21	9–14	3	40	54	58	57	54	53	45	53	68	70	74
2–21	9	0	0	2	25	55	51	49	48	22	9	1	0
2–21	14	0	0	3	22	56	55	51	48	25	8	2	0

Conclusions

1. Based on the experimental results, which factors limited the algae populations? What is your evidence?

2. Based on the experimental results, which factors limited the brine shrimp populations? What is your evidence?

3. What additional abiotic and biotic factors might limit population size in Mono Lake?

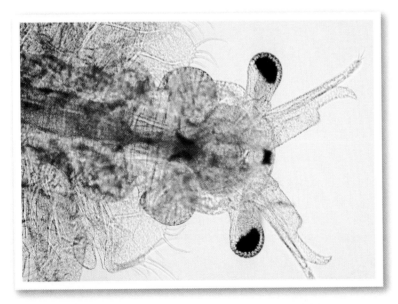

Brine shrimp are tiny crustaceans that are critical to the Mono Lake ecosystem. The black dots are compound eyes set on stalks.

Mono Lake Data

Because of its unique ecology, Mono Lake has been an interesting place to study. A lot is known about the organisms that live in the lake and the abiotic conditions that affect the organisms in the ecosystem.

Good scientific study involves accurate data recording. The first two graphs show how the abiotic factors at Mono Lake change over the course of a year. The other graphs show how some of the populations in the Mono Lake ecosystem change over the course of a year.

Study the graphs. Look for relationships (1) between populations of different species and (2) between organisms and abiotic factors in the ecosystem.

Abiotic Data

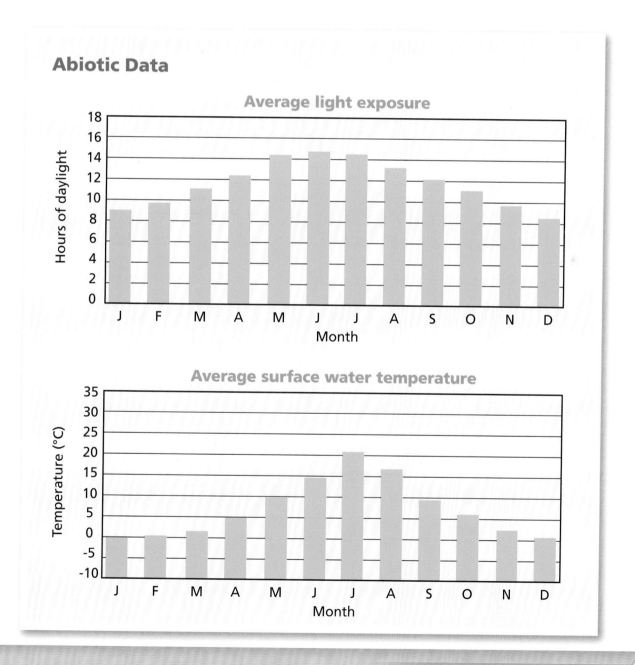

Population Data

Average planktonic algae population

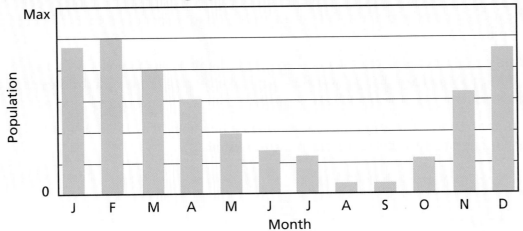

Average brine fly population

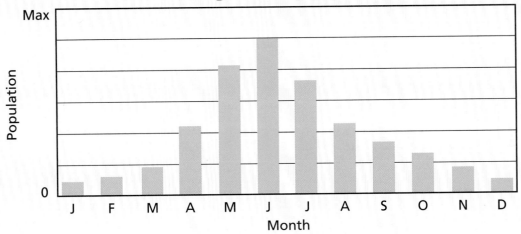

Average brine shrimp population

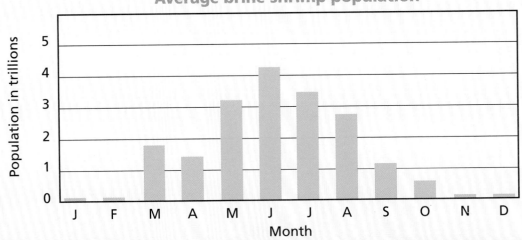

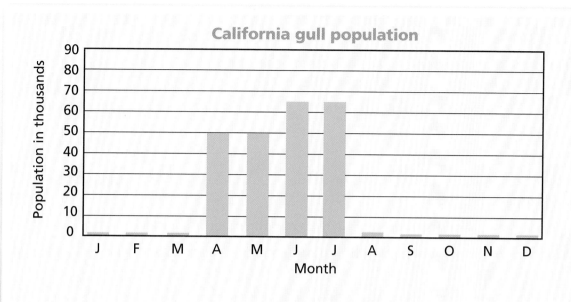

California gull population

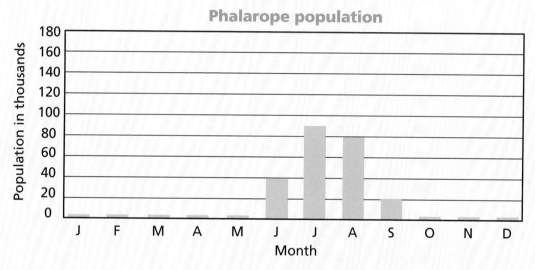

Phalarope population

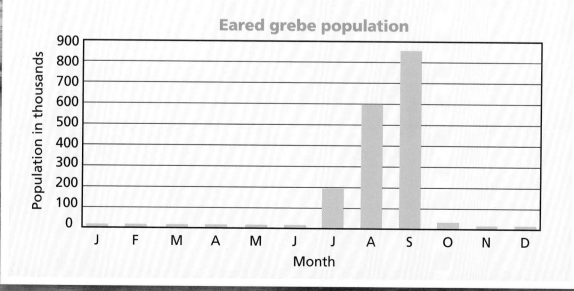

Eared grebe population

Conclusions

1. When does the planktonic algae population peak? When does the brine shrimp population peak? How can you explain the timing of each peak?

2. What is the relationship between water temperature and the brine shrimp and brine flies in the Mono Lake ecosystem?

3. When do the birds arrive?

4. Discuss the similarities and differences in the population graphs of the birds (gulls, phalaropes, and grebes).

5. What is the relationship between the birds and the other organisms?

6. What do you think is going on with the populations at Mono Lake in April?

Science Safety Rules

1. Always follow the safety procedures outlined by your teacher. Follow directions, and ask questions if you're unsure of what to do.

2. Never put any material in your mouth. Do not taste any material or chemical unless your teacher specifically tells you to do so.

3. Do not smell any unknown material. If your teacher asks you to smell a material, wave a hand over it to bring the scent toward your nose.

4. Avoid touching your face, mouth, ears, eyes, or nose while working with chemicals, plants, or animals. Tell your teacher if you have any allergies.

5. Always wash your hands with soap and warm water immediately after using chemicals (including common chemicals, such as salt and dyes) and handling natural materials or organisms.

6. Do not mix unknown chemicals just to see what might happen.

7. Always wear safety goggles when working with liquids, chemicals, and sharp or pointed tools. Tell your teacher if you wear contact lenses.

8. Clean up spills immediately. Report all spills, accidents, and injuries to your teacher.

9. Treat animals with respect, caution, and consideration.

10. Never use the mirror of a microscope to reflect direct sunlight. The bright light can cause permanent eye damage.

Glossary

abiotic nonliving

aerobic cellular respiration a process by which organisms convert sugar into usable energy

anthropocene a new geologic epoch some scientists think we have entered because of human-driven changes

aquatic of the water

atmosphere the thin layer of gases around Earth that extends about 600 kilometers above the surface

atom a particle that is the basic building block of matter

autotroph an organism that makes its own food

biodiversity the variety of life

biodiversity index a measure of biodiversity for an area by dividing the total number of species by the total number of individuals

biomagnification the process by which toxins accumulate in the bodies of predators

biomass the mass of matter produced by organisms in an ecosystem

biome a collection of ecosystems that have similar environments and organisms

biosphere the sum total of all living organisms

biotic consisting of living organisms and products of organisms

calorie a unit used to measure energy transfer

carbohydrate food in the form of sugar or starch

carnivore an organism that eats other animals

carrying capacity the maximum size of a population that a given environment can support

chlorophyll a green pigment in chloroplasts that captures light energy to make sugars during photosynthesis

clutch a cluster of eggs

community all of the interacting populations in a specified area

competition when several kinds of organisms in an ecosystem require the same resources, such as food sources

consumer an organism that eats other organisms

controlled experiment an experiment in which the observer is able to standardize all but one variable to measure results

cultural in the context of ecosystem services, benefits of an ecosystem such as recreation and tourism

decomposer an organism that consumes parts of dead organisms and converts all the biomass into simple chemicals

dehydration when the body doesn't have enough water to function normally

detritivore an organism that eats detritus, breaking the organic material into smaller parts

detritus dead organic material

ecological footprint the size of the impact humans have on an ecosystem

ecosystem a system of interacting organisms and nonliving factors in a specified area

ecosystem engineer a keystone species that changes the ecosystem it inhabits

ecosystem service a benefit that humans obtain from the environment

energy the capacity to do work. Most energy used by organisms comes from the Sun.

environment the surroundings of an organism, including the living and nonliving factors

exoskeleton a tough, outer covering that insects and other organisms have for protection

extinct no longer existing

extinction when the last of a species' population dies out and it can no longer reproduce

food a substance that provides energy and nutrients for organisms

food chain a sequence of organisms that eat one another in an ecosystem

food web all of the feeding relationships in an ecosystem

geosphere the rocky, mineral part of Earth

habitat a place where an organism lives that supports its requirements for life

herbivore an organism that eats only plants

heterotroph an organism that cannot make its own food and must eat other organisms

hydrosphere all of the water on Earth including the ocean, lakes, rivers, streams, aquifers, polar icecaps, glaciers, snowpacks, permafrost, and condensation

incomplete metamorphosis the process of gradual maturing of an insect through stages (egg, nymphal stages or instars, adult)

individual a single organism

inference an explanation or assumption that people make based on their knowledge, experiences, or opinions

instar an immature nymphal stage of an insect as it grows into an adult form

interdependent reliant on each other

introduced species a species that is brought to a place by humans and becomes part of the local ecosystem

invasive species an introduced species that has a negative impact on an ecosystem

keystone species a species that is critical for the overall health of an ecosystem

limiting factor any biotic or abiotic component of the ecosystem that controls the size of a population

mass the amount of matter in something

migrate to move from one area to another according to seasonal changes

model a representation of thinking to help others understand your thinking

molecule a particle made of two or more smaller particles held together by chemical bonds

molting the process of shedding an exoskeleton in order to grow

native species a kind of organism that has been part of an ecosystem for a long time

niche a role within an ecosystem

nymph an immature bug

observation noticing the properties of an object or event with one or more of the five senses (sight, hearing, touch, smell, and taste)

observational study an experiment in which the observer collects data over time without interacting with the area of study

omnivore a consumer that eats both plants and animals

organism a living thing

pH a measurement of how acidic or basic a liquid is

photosynthesis the process by which producers make energy-rich molecules (food) from water and carbon dioxide in the presence of light

phytoplankton a huge array of photosynthetic microorganisms that are free-floating in water

polar zone the climate zone that is closest to the North and South Poles (latitude 60°–90° north and latitude 60°–90° south)

population all the individuals of one kind (one species) in a specified area at one time

population study an experiment in which the observer collects data over time for one population in an ecosystem

predator an organism, usually an animal, that eats other organisms

prey an organism, usually an animal, that is eaten by another organism

primary consumer an organism that eat producers; also called first-level consumer

primary production the amount of food produced and redistributed throughout the food web during a year

proboscis a tube-like beak for sucking fluids from plants. True bugs have this structure.

producer an organism that is able to produce its own food through photosynthesis

protist a eukaryotic organism that is not an animal, plant, or fungus; includes algae

provisioning in the context of ecosystem services, benefits of an ecosystem such as food, water, or fuel

regulating in the context of ecosystem services, benefits of an ecosystem such as climate regulation or disease control

reproductive potential the theoretical unlimited growth of a population over time

sampling a technique to count the organisms in a selected area to make inferences about the total number of organisms

secondary consumer an organism that eats primary consumers; also called second-level consumer

semiarid very dry, but not completely dry

species a kind of organism. Members of a species are all the same kind of organism and are different from all other kinds of organisms.

supporting in the context of ecosystem services, benefits of an ecosystem that support other services, such as water cycling and soil formation

sustain to last or maintain over a long period of time

sustainable can last or maintain itself over a long period of time

symbiotic a relationship between species in which one or both species rely on each other; one or both species benefit from the relationship

temperate zone the climate zone lying between the Tropic of Cancer and the Arctic Circle in the Northern Hemisphere or between the Tropic of Capricorn and the Antarctic Circle in the Southern Hemisphere

terrestrial of the land

tertiary consumer an organism that eats secondary consumers; also called third-level consumer

thermal energy radiant energy that heats

trophic level a functioning role in a feeding relationship through which energy flows

trophic pyramid a trophic-level diagram in which the largest layer at the base is the producers; the first-level, second-level, and third-level consumers are in the layers above

tropical zone the climate zone closest to the equator (latitude 30° south to 30° north)

tufa tower a naturally occurring, gray, lumpy structure that forms underwater in a salt lake because of a chemical reaction between calcium and salt

zooplankton microscopic adult animals and larval forms of animals found free-floating in fresh water and sea water

All tigers belong to the same species of big cats, *Panthera tigris*. Only six subspecies of tigers remain, including this one, the Siberian tiger.

Index